# Inference for Heavy-Tailed Data

# INFERENCE FOR HEAVY-TAILED DATA

## Applications in Insurance and Finance

**Liang Peng**
Georgia State University
Department of Risk Management and Insurance
Robinson College of Business
Atlanta, GA 30303, USA

**Yongcheng Qi**
University of Minnesota – Duluth
Department of Mathematics and Statistics
1117 University Drive
Duluth, MN 55812, USA

ELSEVIER

ACADEMIC PRESS
An imprint of Elsevier

Academic Press is an imprint of Elsevier
125 London Wall, London EC2Y 5AS, United Kingdom
525 B Street, Suite 1800, San Diego, CA 92101-4495, United States
50 Hampshire Street, 5th Floor, Cambridge, MA 02139, United States
The Boulevard, Langford Lane, Kidlington, Oxford OX5 1GB, United Kingdom



**Library of Congress Cataloging-in-Publication Data**
A catalog record for this book is available from the Library of Congress

**British Library Cataloguing-in-Publication Data**
A catalogue record for this book is available from the British Library

ISBN: 978-0-12-804676-0



*Publisher:* Candice Janco
*Acquisition Editor:* Glyn Jones
*Editorial Project Manager:* Lindsay Lawrence
*Production Project Manager:* Omer Mukthar
*Designer:* Matthew Limbert

Typeset by VTeX

# CONTENTS

# ABOUT THE AUTHORS

**Liang Peng** is Thomas Bowles Professor of Actuarial Science in the Department of Risk Management and Insurance at Georgia State University. He obtained his Ph.D. from Erasmus University Rotterdam in the Netherlands. So far he has written more than 130 articles on extreme value theory, empirical likelihood methods, time series analysis and risk analysis.

**Yongcheng Qi** is Professor of Statistics in the Department of Mathematics and Statistics at University of Minnesota – Duluth. He obtained his Ph.D. from Peking University in China and University of Georgia in USA. So far he has written more than 87 articles on extreme value theory, probability theory, and nonparametric statistics.

# PREFACE

Heavy tailed data frequently appear in insurance and finance, and a loss variable with a heavy tail rarely creates unusually huge losses. Unfortunately such an extreme loss often causes severe damages to our society. Extreme value theory has been developed to model, analyze and predict such an extreme event for decades.

Several excellent books on extreme value theory have been available in the market such as Leadbetter et al. [67], Resnick [91], Embrechts et al. [36], Coles [22], Beirlant et al. [5], de Haan and Ferreira [27] and Novak [75]. With little overlapping with these existing books, this short book aims to collect some recent statistical inference methods for analyzing heavy tailed data. This collection heavily relies on our own research experience and understanding of difficulties in applying extreme value theory to real life data in insurance and finance, and so we surely miss many other good methods.

Chapter 1 collects some definitions and notations in probability theory and extreme value theory. Chapter 2 uses a well-known tail index estimator to address issues and methods for analyzing heavy tailed independent data such as the applications of tail empirical process, tail quantile process and inequalities for a regularly varying function, the choice of sample fraction in tail index estimation and high quantile estimation, goodness-of-fit tests for heavy tailed distribution functions, and expected shortfall with a possible infinite variance loss. Chapter 3 collects some methods for analyzing heavy tailed dependent data with a focus on time series models such as ARMA models and GARCH models. Chapter 4 collects some applications of multivariate regular variation in risk analysis. Finally the collected inference procedures are applied to some real data sets in insurance and finance in Chapter 5.

Writing such a short book is not easy unlike we thought and planned in the beginning. Selecting topics and unifying notations are quite time-consuming. We are grateful to our families for their support. Without their understanding, we can not sacrifice so much of our family time to finish this book on time.

We also thank Lindsay Lawrence, Glyn Jones and the team at Elsevier for guidance and help throughout the publishing process.

Liang Peng, Atlanta, Georgia
Yongcheng Qi, Duluth, Minnesota
May 2017

# CHAPTER 1

# Introduction

This chapter collects some useful definitions and notations in probability theory and extreme value theory.

## 1.1 BASIC PROBABILITY THEORY

Let $\Omega$ be a space, which is an arbitrary, nonempty set. Write $\omega \in \Omega$ if $\omega$ is an element of $\Omega$, and write $A \subseteq \Omega$ if $A$ is a subset of $\Omega$.

**Definition 1.1.** A nonempty class $\mathcal{A}$ of subsets of $\Omega$ is called an **algebra** if

**i)** the complementary set $A^c \in \mathcal{A}$ whenever $A \in \mathcal{A}$; and

**ii)** the union $A_1 \cup A_2 \in \mathcal{A}$ whenever $A_1 \in \mathcal{A}$ and $A_2 \in \mathcal{A}$.

Moreover, $\mathcal{A}$ is called a $\sigma$**-algebra** or a $\sigma$**-field** if, in addition to i) and ii),

**iii)** $\cup_{i=1}^{\infty} A_i \in \mathcal{A}$ whenever $A_i \in \mathcal{A}$ for $i \geq 1$.

**Definition 1.2.** If $\mathcal{A}$ is a $\sigma$-algebra with respect to the space $\Omega$, then the pair $(\Omega, \mathcal{A})$ is called a **measurable space**. The sets of $\mathcal{A}$ are called **measurable sets**.

**Definition 1.3.** The elements of the $\sigma$-algebra $\mathcal{B}$ generated by the class of infinite intervals of the form $[-\infty, x)$ for $-\infty < x < \infty$ are called **Borel sets**. The measurable space $(\bar{\mathbb{R}} = [-\infty, \infty], \mathcal{B})$ is called **Borel space**.

**Definition 1.4.** If $(\Omega_1, \mathcal{A}_1)$ and $(\Omega_2, \mathcal{A}_2)$ are two measurable spaces and $f$ is a mapping from $\Omega_1$ to $\Omega_2$, then $f$ is said to be a **measurable transformation/mapping** if $f^{-1}(A) \in \mathcal{A}_1$ for any $A \in \mathcal{A}_2$, where $f^{-1}(A) = \{\omega : \omega \in \Omega_1, f(\omega) \in A\}$.

**Definition 1.5.** For a measurable space $(\Omega, \mathcal{A})$, a set function $P$ defined on $\mathcal{A}$ is called a **probability** if

**i)** $P(\emptyset) = 0$, where $\emptyset$ denotes the empty set;

**ii)** $P(A \cup B) = P(A) + P(B)$ for disjoint events $A, B \in \mathcal{A}$ (i.e., $A \cap B = \emptyset$);

**iii)** $P(\cup_{i=1}^{\infty} A_i) = \sum_{i=1}^{\infty} P(A_i)$ for disjoint events $A_i \in \mathcal{A}$, $i = 1, 2, \cdots$.

In this case, $(\Omega, \mathcal{A}, P)$ is called a **probability space**.

**Definition 1.6.** A real-valued measurable function $X$ on a probability space $(\Omega, \mathcal{A}, P)$ is called a **random variable**. The function

$$F(x) = P(X \leq x) := P(\{\omega \in \Omega : X(\omega) \leq x\}) \quad \text{for} \quad x \in \mathbb{R} = (-\infty, \infty)$$

**Inference for Heavy-Tailed Data.**
**DOI: http://dx.doi.org/10.1016/B978-0-12-804676-0.00001-8**

is called the **cumulative distribution function** or **distribution function** of $X$.

**Definition 1.7.** A sequence of random variables $\{X_n\}_{n=1}^{\infty}$ defined on a probability space $(\Omega, \mathcal{A}, P)$ is called **independent** if for any $m \geq 1$, $1 \leq i_1 < \cdots < i_m < \infty$ and $-\infty < x_1, \cdots, x_m < \infty$

$$P(X_{i_1} \leq x_1, \cdots, X_{i_m} \leq x_m) = \prod_{j=1}^{m} P(X_{i_j} \leq x_j).$$

**Definition 1.8.** If $\{X_n\}_{n=0}^{\infty}$ is a sequence of random variables on a probability space $(\Omega, \mathcal{A}, P)$, then $\{X_n\}_{n=1}^{\infty}$ is said to **converge in probability** to $X_0$ (notation: $X_n \xrightarrow{p} X_0$) if for any $\epsilon > 0$

$$\lim_{n\to\infty} P(|X_n - X_0| > \epsilon) = 0.$$

**Definition 1.9.** If $\{X_n\}_{n=0}^{\infty}$ is a sequence of random variables on a probability space $(\Omega, \mathcal{A}, P)$ with corresponding cumulative distribution functions $\{F_n(x)\}_{n=0}^{\infty}$, then $\{X_n\}_{n=1}^{\infty}$ is said to **converge in distribution** to $X_0$ (notation: $X_n \xrightarrow{d} X_0$ or $X_n \xrightarrow{d} F_0$) if for any continuity point $x$ of $F_0$

$$\lim_{n\to\infty} F_n(x) = F_0(x).$$

**Definition 1.10.** A sequence of random variables $\{X_n\}$ on a probability space $(\Omega, \mathcal{A}, P)$ is said to be **bounded in probability** if for any $\epsilon > 0$, there exist constants $C > 0$ and integer $N$ such that

$$P(|X_n| > C) \leq \epsilon \quad \text{for all} \quad n \geq N.$$

Let $\{X_n\}$ be a sequence of random variables on a probability space $(\Omega, \mathcal{A}, P)$ and $\{b_n\}$ be a sequence of positive constants. We write $X_n = o_p(b_n)$ if $X_n/b_n \xrightarrow{p} 0$, and write $X_n = O_p(b_n)$ if $X_n/b_n$ is bounded in probability.

**Definition 1.11.** A **stochastic process** is a collection $\{X_t : t \in T\}$, where $T$ is a subset of $\mathbb{R}$ and $X_t$ is a random variable on a probability space $(\Omega, \mathcal{A}, P)$.

**Definition 1.12.** A **Wiener process** $\{W(t) : t \geq 0\}$ is a continuous-time stochastic process satisfying

**i)** $W(0) = 0$;

**ii)** $W(t+u) - W(t)$ is independent of the $\sigma$-algebra generated by $\{W(s) : 0 < s \leq t\}$ for any $u > 0$;

**iii)** $W(t+u) - W(t)$ has a normal distribution with mean zero and variance $u$ for any $u > 0$.

**Definition 1.13.** If $W(t)$ for $t \geq 0$ is a Wiener process, then $B(t) = W(t) - \frac{t}{T}W(T)$ is called a **Brownian Bridge** on $[0, T]$. In this case,

$$B(0) = B(T) = 0 \quad \text{and} \quad E\big(B(s)B(t)\big) = s(T-t)$$

for $0 \leq s < t \leq T$, but the increments are no longer independent.

**Definition 1.14.** The space $D[0, 1]$ denotes the space of functions on $[0, 1]$ that are right continuous and have left-hand limits.

For the space $(E, \varepsilon)$, let $C_K(E)$ be the set of all continuous, real valued functions on $E$ with compact support, and $C_K^+(E)$ be the subset of $C_K(E)$ consisting of continuous, nonnegative functions with compact support. Let $M_+(E)$ be the set of all nonnegative Radon measures on $(E, \varepsilon)$ and define $\mu_+(E)$ to be the smallest $\sigma$-field of subsets of $M_+(E)$ making the maps $m \to m(f) = \int_E f\, dm$ from $M_+(E) \to \mathbb{R}$ measurable for all $f \in C_K^+(E)$. Here **Radon** means the measure of compact sets is always finite.

**Definition 1.15.** $\xi$ is a **random measure** if it is a measurable map from a probability space $(\Omega, \mathcal{A}, P)$ into $(M_+(E), \mu_+(E))$.

**Definition 1.16.** For $\mu_n, \mu \in M_+(E)$, we say $\mu_n$ **converges vaguely** to $\mu$ (written $\mu_n \xrightarrow{v} \mu$) if $\mu_n(f) \to \mu(f)$ for all $f \in C_K^+(E)$.

**Definition 1.17.** $C \subset \mathbb{R}^d$ is a **cone** if $t\boldsymbol{x} \in C$ for every $t > 0$ and $\boldsymbol{x} \in C$.

Let $\Lambda$ denote the class of strictly increasing, continuous mappings of $[0, 1]$ onto itself with $\lambda(0) = 0$ and $\lambda(1) = 1$ for each $\lambda \in \Lambda$. Given $x$ and $y$ in the space $D[0, 1]$, define $d(x, y)$ to be the infimum of those positive $\epsilon$ for which there exists a $\lambda \in \Lambda$ such that

$$\sup_t |\lambda(t) - t| \leq \epsilon \quad \text{and} \quad \sup_t |x(t) - y(\lambda(t))| \leq \epsilon.$$

In this way, $d(x, y)$ defines the **Skorohod topology**.

**Definition 1.18.** Let $F$ be the cumulative distribution function of a random variable $X$. Then the **generalized inverse** of $F$ is defined as

$$F^{-}(u) = \inf\{t : F(t) \geq u\} \quad \text{for} \quad 0 < u < 1. \tag{1.1}$$

**Lemma 1.1.** *Let $F$ be a cumulative distribution function.*

**i)** *For any $x \in \mathbb{R}$ and $u \in (0,1)$, $F^{-}(u) \le x$ if and only if $u \le F(x)$.*

**ii)** *If $U$ is a random variable with uniform distribution over $(0,1)$, then the distribution function of $F^{-}(U)$ is $F(x)$.*

**iii)** *If $F$ is continuous, then $F(F^{-}(u)) = u$ for $0 < u < 1$.*

*Proof.* i) If $u \le F(x)$, then $x \in \{t : F(t) \ge u\}$, which implies that

$$x \ge \inf\{t : F(t) \ge u\} = F^{-}(u).$$

Next assume that $F^{-}(u) \le x$. Since $F$ is right continuous, we have

$$F(x) \ge F(F^{-}(u)) = F(\inf\{t : F(t) \ge u\}) = \inf\{F(t) : F(t) \ge u\} \ge u.$$

Hence part i) follows.

ii) It follows from part i) that $P(F^{-}(U) \le x) = P(U \le F(x)) = F(x)$ for any $x$. That is, $F^{-}(U)$ has the distribution function $F(x)$.

iii) It follows obviously. □

Like Csörgő et al. [25], we use the following conventions concerning integrals.

When $a < b$ and $g$ is a left-continuous and $f$ is a right-continuous function, then

$$\int_a^b f\,dg = \int_{[a,b)} f\,dg \quad \text{and} \quad \int_a^b g\,df = \int_{(a,b]} g\,df, \tag{1.2}$$

whenever these integrals make sense as Lebesgue–Stieltjes integrals. In this case the usual integration by parts formula

$$\int_a^b f\,dg + \int_a^b g\,df = g(b)f(b) - g(a)f(a) \tag{1.3}$$

is valid.

For any Brownian bridge $\{B(s) : 0 \le s \le 1\}$, and with $0 \le a < b \le 1$ and the functions $f$ and $g$ as above we define the following stochastic integral

$$\int_a^b f(s)\,dB(s) = f(b)B(b) - f(a)B(a) - \int_a^b B(s)\,df(s) \tag{1.4}$$

and the same formula for $g$ replacing $f$.

## 1.2 BASIC EXTREME VALUE THEORY

Let $X_1, \cdots, X_n$ be a random sample of size $n$ from a distribution function $F$, that is, $X_1, \cdots, X_n$ are independent and identically distributed (i.i.d.) random variables with distribution function $F$. The **univariate extreme value theory** is based on the assumption that there exist constants $a_n > 0$ and $b_n \in \mathbb{R}$ such that

$$\frac{\max_{1\le j\le n} X_j - b_n}{a_n} \xrightarrow{d} G \quad \text{as } n \to \infty, \tag{1.5}$$

where $G$ is a non-degenerate distribution function. In this case $G$ is called an **extreme value distribution** and $F$ is said to be in the **domain of (maximum) attraction** of the extreme value distribution $G$ (notation: $F \in D(G)$). To classify $G$, we need the following definition.

**Definition 1.19.** Two distribution functions $F(x)$ and $G(x)$ are said to have the **same type** if for some constants $a > 0$ and $b \in \mathbb{R}$

$$G(x) = F(ax + b) \quad \text{for all } x \in \mathbb{R}.$$

**Lemma 1.2.** *Let $\{X_n\}$, $U$, $V$ be random variables such that neither $U$ nor $V$ is degenerate (i.e., both $U$ and $V$ are non-constant). If there are constants $a_n > 0$, $\alpha_n > 0$, $b_n \in \mathbb{R}$, $\beta_n \in \mathbb{R}$ such that*

$$\frac{X_n - b_n}{a_n} \xrightarrow{d} U \quad \text{and} \quad \frac{X_n - \beta_n}{\alpha_n} \xrightarrow{d} V \quad \text{as} \quad n \to \infty,$$

*then*

$$\lim_{n\to\infty} \frac{\alpha_n}{a_n} = A > 0, \quad \lim_{n\to\infty} \frac{\beta_n - b_n}{\alpha_n} = B \in \mathbb{R} \quad \text{and} \quad V \stackrel{d}{=} \frac{U - B}{A}.$$

*Proof.* See Proposition 0.2 of Resnick [91]. □

Using the above notion of the same type, it is well-known that the limiting distribution $G$ in (1.5) must be one of the following three types:

- Reversed Weibull distribution

$$\Phi_\alpha(x) = \begin{cases} \exp(-(-x)^\alpha), & x < 0, \\ 1, & x \ge 0, \end{cases}$$

where $\alpha > 0$;

- Gumbel distribution

$$\Lambda(x) = \exp(-e^{-x}), \quad x \in \mathbb{R};$$

- Fréchet distribution

$$\Psi_\alpha(x) = \begin{cases} 0, & x \le 0, \\ \exp(-x^{-\alpha}), & x > 0, \end{cases}$$

where $\alpha > 0$.

A unified expression for $G$ in (1.5) is

$$G_\gamma(x) = \exp\{-(1+\gamma x)^{-1/\gamma}\}, \quad \text{where} \quad 1+\gamma x > 0. \tag{1.6}$$

Here $\gamma \in \mathbb{R}$ is called the **extreme value index**, and $\gamma < 0, \gamma = 0, \gamma > 0$ correspond to the reversed Weibull distribution, Gumbel distribution, Fréchet distribution, respectively. For modeling losses in insurance and finance, this book focuses on the case of $\gamma > 0$, i.e., the Fréchet distribution in (1.5).

For the study of extreme comovement of financial markets, multivariate extreme value theory is needed, which is based on the assumption that there exist constant vectors $\mathbf{a}_n > \mathbf{0}, \mathbf{b}_n \in \mathbb{R}^d$ and a non-degenerate $d$-dimensional distribution function $H$ such that

$$\frac{\max_{1\le j\le n} \mathbf{Z}_j - \mathbf{b}_n}{\mathbf{a}_n} \xrightarrow{d} H \quad \text{as } n \to \infty, \tag{1.7}$$

where $\{\mathbf{Z}_i = (Z_{i1}, \cdots, Z_{id})^T, i \ge 1\}$ is a sequence of i.i.d. random vectors in $\mathbb{R}^d$ with a common distribution function $F(z_1, \cdots, z_d)$ and marginal distributions $F_j(z_j)$ for $j = 1, \cdots, d$. Throughout we use $A^T$ to denote the transpose of the vector or matrix $A$.

Like the univariate extreme value theory, $H$ in (1.7) is called a multivariate extreme value distribution and $F$ is said to be in the domain of attraction of $H$ (notation: $F \in D(H)$). Since the convergence of the joint distributions for multivariate extremes implies the convergence of the marginal distributions, the marginal distribution $H_i$ of $H$ must be one of the three types of extreme value distributions. Also $H$ is a continuous function since its marginal distributions are continuous.

Also note that (1.7) is equivalent to

$$\lim_{n\to\infty} F^n(a_{n1}z_1 + b_{n1}, \cdots, a_{nd}z_d + b_{nd}) = H(z_1, \cdots, z_d) \tag{1.8}$$

for all $(z_1, \cdots, z_d) \in \mathbb{R}^d$, which is equivalent to

$$\lim_{n\to\infty} n\{1 - F(a_{n1}z_1 + b_{n1}, \cdots, a_{nd}z_d + b_{nd})\} = -\log H(z_1, \cdots, z_d)$$

for all $(z_1, \cdots, z_d)$ such that $0 < H(z_1, \cdots, z_d) < 1$. This can further be decomposed as the following marginal conditions

$$F_j \in D(H_{\gamma_j}) \text{ with } H_{\gamma_j} \text{ given in (1.6) and } \gamma_j \in \mathbb{R} \tag{1.9}$$

for $j = 1, \cdots, d$, and the following dependence condition

$$\begin{aligned}\lim_{n\to\infty} n\{1 - F(U_1(nx_1), \cdots, U_d(nx_d))\} &= -\log H(\frac{x_1^{\gamma_1}-1}{\gamma_1}, \cdots, \frac{x_d^{\gamma_d}-1}{\gamma_d}) \\ &=: l(x_1, \cdots, x_d),\end{aligned} \tag{1.10}$$

where $U_i(x) = F_i^-(1 - 1/x)$ for $i = 1, \cdots, d$. The limiting function $l(x_1, \cdots, x_n)$ in (1.10) is called a **tail dependence function**, which is a homogeneous function satisfying

$$l(tx_1, \cdots, tx_d) = t^{-1} l(x_1, \cdots, x_d) \quad \text{for any } t > 0.$$

This homogeneous tail dependence function plays an important role in extrapolating multivariate data into a far tail region for predicting an extreme event.

# CHAPTER 2

# Heavy Tailed Independent Data

This chapter collects some known statistical inference methods for analyzing univariate independent data with a common distribution function in the domain of attraction of an extreme value distribution with a positive extreme value index, i.e., (1.5) and (1.6) hold with some $\gamma > 0$.

## 2.1 HEAVY TAIL

**Definition 2.1.** A distribution function $F(x)$ is said to have a **(right) heavy tail** with tail index $\alpha > 0$ if it satisfies that

$$\lim_{t\to\infty} \frac{1 - F(tx)}{1 - F(t)} = x^{-\alpha} \quad \text{for all} \quad x > 0. \tag{2.1}$$

Likewise, a random variable $X$ with such a distribution function $F$ in (2.1) is called a heavy tailed random variable.

**Definition 2.2.** A measurable function $a(x)$ defined over $(0, x_0)$ for some $x_0 > 0$ is said to be **regularly varying** or a **regular variation** at zero with an exponent $\rho \in \mathbb{R}$ (notation $a(x) \in RV_\rho^0$) if

$$\lim_{t\to 0} \frac{a(tx)}{a(t)} = x^\rho \quad \text{for all} \quad x > 0.$$

It is true that $a(x) \in RV_\rho^0$ if and only if $a(1/x) \in RV_{-\rho}^\infty$. Hence, when (2.1) holds, the survival function, $\bar{F}(x) := 1 - F(x)$, is also said to be a regularly varying function at infinity with index $-\alpha$. Moreover, if (2.1) holds with $\alpha = 0$, then $\bar{F}(x)$ is called a **slowly varying** function at infinity.

It is known that $F \in D(G_\gamma)$ with some $\gamma > 0$ is equivalent to $\bar{F} \in RV_{-1/\gamma}^\infty$, where $G_\gamma$ is given in (1.6).

Put $x_+ = \max(x, 0)$. When the distribution function of a random variable $X$ satisfies (2.1), we have that $E(X_+^\gamma) = \infty$ for any $\gamma > \alpha$ and $E(X_+^\gamma) < \infty$ for any $0 < \gamma < \alpha$. Some heavy tailed distribution functions include:

- *Fréchet distribution:* $\bar{F}(x) = 1 - \exp\left\{-x^{-\alpha}\right\}$ for $x > 0$, where $\alpha > 0$. The tail index is $\alpha$.
- *Pareto distribution:* $\bar{F}(x) = x^{-\alpha}$ for $x \geq 1$, where $\alpha > 0$. The tail index is $\alpha$ (notation: Pareto($\alpha$)).

**Inference for Heavy-Tailed Data.**
**DOI: http://dx.doi.org/10.1016/B978-0-12-804676-0.00002-X** 

- *Cauchy distribution:* $\bar{F}(x) = \int_x^\infty \frac{1}{\pi(1+t^2)}\,dt$ for $-\infty < x < \infty$. The tail index is 1.
- *t-distribution:* $\bar{F}(x) = \int_x^\infty \frac{\Gamma((\nu+1)/2)}{\Gamma(\nu/2)\sqrt{\nu\pi}}(1+\frac{t^2}{\nu})^{-\frac{\nu+1}{2}}\,dt$ for $-\infty < x < \infty$, where $\nu > 0$. The tail index is $\nu$.
- *Burr distribution:* $\bar{F}(x) = (1 + x^b)^{-a}$ for $x \geq 0$, where $a, b > 0$. The tail index is $ab$ (notation: Burr$(a, b)$).
- *Log-Gamma distribution:* $\bar{F}(x) = \int_x^\infty \frac{\alpha^\beta}{\Gamma(\beta)}(\log t)^{\beta-1}t^{-\alpha-1}\,dt$ for $x \geq 1$, where $\alpha > 0, \beta > 0$. The tail index is $\alpha$.
- Assume that $\xi$ and $\eta$ are two independent random variables, $\eta$ is a heavy tailed random variable, and $\xi \geq 0$. Then $\xi\eta$ can be a heavy tailed random variable according to the following Breiman's lemma.

  *Breiman's Lemma:* the random variable $\xi\eta$ has a heavy tail with index $\alpha$ and satisfies

  $$\lim_{t\to\infty} \frac{P(\xi\eta > t)}{P(\eta > t)} = E(\xi^\alpha) \quad \text{if} \quad \lim_{t\to\infty} \frac{P(\eta > tx)}{P(\eta > t)} = x^{-\alpha} \quad \text{for all} \quad x > 0$$

  and the nonnegative random variable $\xi$ satisfies $E\xi^\gamma < \infty$ for some $\gamma > \alpha > 0$.

For studying the asymptotic behavior of extremes, it is often convenient to write (2.1) in terms of its inverse function. Indeed, (2.1) is equivalent to

$$\lim_{t\to 0} \frac{\bar{F}^-(tx)}{\bar{F}^-(t)} = x^{-1/\alpha} \quad \text{for all} \quad x > 0, \tag{2.2}$$

which is equivalent to

$$\lim_{t\to 0} \frac{(tx)^{1/\alpha}\bar{F}^-(tx)}{t^{1/\alpha}\bar{F}^-(t)} = 1 \quad \text{for all} \quad x > 0, \tag{2.3}$$

where $\bar{F}^-(x) = F^-(1 - x)$ for $0 < x < 1$, and $F^-$ denotes the generalized inverse function of $F$ as defined in (1.1). Hence condition (2.1) (i.e., $\bar{F}(x) \in RV_{-\alpha}^\infty$) is equivalent to $\bar{F}^-(x) \in RV_{-1/\alpha}^0$, and is equivalent to $x^{1/\alpha}\bar{F}^-(x) \in RV_0^0$.

In order to specify an approximation rate in (2.2) or (2.3), which plays an important role in deriving the asymptotic properties of estimators for the tail index $\alpha$ and some related quantities such as a high quantile and an extreme tail probability, one could assume that there exist a function $c(x) \not\equiv 0$ and a function $A(t) \to 0$ with a constant sign near zero such that

$$\lim_{t\to 0} \frac{(tx)^{1/\alpha}\bar{F}^-(tx) - t^{1/\alpha}\bar{F}^-(t)}{A(t)} = c(x) \quad \text{for all} \quad x > 0.$$

In this case, $t^{1/\alpha}\bar{F}^{-}(t)$ is called a ***Π*-variation**, and by Theorem B.2.1 of de Haan and Ferreira [27], we could assume that there exist some $\rho \geq 0$ and a function $A(t) \in RV_{\rho}^{0}$ with $\lim_{t\to 0} A(t) = 0$ such that

$$\lim_{t\to 0} \frac{(tx)^{1/\alpha}\bar{F}^{-}(tx) - t^{1/\alpha}\bar{F}^{-}(t)}{A(t)} = \frac{x^{\rho}-1}{\rho} \quad \text{for all} \quad x > 0. \tag{2.4}$$

When we discuss bias corrected tail index estimation later, we need to further specify an approximation rate in (2.4). By de Haan and Stadtmüller [28], we could generally assume that there exist some $\gamma \geq 0$ and a function $B(t) \in RV_{\gamma}^{0}$ with $\lim_{t\to 0} B(t) = 0$ such that

$$\begin{aligned} &\lim_{t\to 0} \frac{\frac{(tx)^{1/\alpha}\bar{F}^{-}(tx) - t^{1/\alpha}\bar{F}^{-}(t)}{A(t)} - \frac{x^{\rho}-1}{\rho}}{B(t)} \\ &= \frac{1}{\gamma}\left(\frac{x^{\rho+\gamma}-1}{\rho+\gamma} - \frac{x^{\rho}-1}{\rho}\right) := H_{\rho,\gamma}(x) \quad \text{for all} \quad x > 0. \end{aligned} \tag{2.5}$$

Here condition (2.5) is also called a **second order regular variation** condition for the function $t^{1/\alpha}\bar{F}^{-}(t)$.

**Example 2.1.** Suppose $1 - F(x) = Cx^{-\alpha}\left\{1 + Dx^{-\rho\alpha} + o(x^{-\rho\alpha})\right\}$ for some $\alpha > 0, C > 0, D \neq 0, \rho > 0$ as $x \to \infty$. Then

$$\bar{F}^{-}(t) = C^{1/\alpha}t^{-1/\alpha}\left\{1 + \alpha^{-1}DC^{-\rho}t^{\rho} + o(t^{\rho})\right\} \quad \text{as} \quad t \to 0,$$

i.e., (2.4) holds with $A(t) = \rho\alpha^{-1}DC^{1/\alpha-\rho}t^{\rho}$.

Although (2.2), (2.4) and (2.5) are defined for each fixed $x > 0$, they do have a sort of uniform convergence property as demonstrated by the following inequalities. This type of uniform convergence plays a useful role in deriving the asymptotic behavior of estimators and tests in analyzing extremes.

- *Potter's bound* (Bingham et al. [9]). Assume $f(x) \in RV_{\rho}^{0}$ for some $\rho \in \mathbb{R}$. Then for any $\epsilon > 0$ and $\delta > 0$, there exists $t_0 > 0$ such that for any $0 < t \leq t_0$ and $0 < tx \leq t_0$,

$$(1-\epsilon)x^{\rho}e^{-\delta|\log x|} \leq \frac{f(tx)}{f(t)} \leq (1+\epsilon)x^{\rho}e^{\delta|\log x|} \tag{2.6}$$

and

$$\left|\frac{f(tx)}{f(t)} - x^{\rho}\right| \leq \epsilon x^{\rho}e^{\delta|\log x|}. \tag{2.7}$$

*Proof.* By the representation theorem of a regular variation, there exist a $t_1 > 0$, functions $b(t)$ and $c(t)$ with

$$\lim_{t\to 0} c(t) = c_0 \in (0,\infty) \quad \text{and} \quad \lim_{t\to 0} b(t) = -\rho$$

such that for all $0 < t < t_1$

$$f(t) = c(t)\exp\left\{\int_t^{t_1} \frac{b(s)}{s}\,ds\right\}.$$

Hence

$$\frac{f(tx)}{f(t)} = \frac{c(tx)}{c(t)}\exp\left\{\int_{tx}^{t} \frac{b(s)}{s}\,ds\right\},$$

which can be used to prove (2.6) and (2.7) straightforwardly. □

- *Inequality for Π-variation* (Theorem B.2.18 of de Haan and Ferreira [27]). Assume that $f(x)$ is a measurable function defined on $(0, x_0)$ for some $x_0 > 0$, and there exist some $\rho \geq 0$ and a function $A(t) \in RV_\rho^0$ with $\lim_{t\to 0} A(t) = 0$ such that

$$\lim_{t\to 0}\frac{f(tx) - f(t)}{A(t)} = \frac{x^\rho - 1}{\rho} \quad \text{for all} \quad x > 0. \tag{2.8}$$

Then for any $\epsilon > 0$ and $\delta > 0$ there is $t_0 \in (0,1)$ such that for all $0 < t \leq t_0$ and $0 < tx \leq t_0$

$$\left|\frac{f(tx) - f(t)}{A(t)} - \frac{x^\rho - 1}{\rho}\right| \leq \epsilon\left\{1 + x^\rho e^{\delta|\log x|}\right\}. \tag{2.9}$$

*Proof.* When $\rho > 0$, we have $f(0) := \lim_{t\to 0} f(t)$ is finite, $a(t) := f(t) - f(0) \in RV_\rho^0$ and $a(t)/A(t) \to \rho^{-1}$ as $t \to 0$, which imply (2.9) by using (2.7) and writing that

$$\frac{f(tx) - f(t)}{A(t)} = \frac{\frac{a(tx)}{a(t)} - x^\rho}{A(t)/a(t)} + \frac{x^\rho - 1}{A(t)/a(t)}.$$

When $\rho = 0$, for a $t_1 > 0$, define $a(t) = f(t) - t^{-1}\int_t^{t_1} f(s)\,ds$. Then we have

$$\frac{a(t)}{A(t)} \to 1,\ a(t) \in RV_0^0,\ \text{and}\ f(t) = a(t) + \int_t^{t_1}\frac{a(s)}{s}\,ds. \tag{2.10}$$

Write

$$\begin{aligned}&\frac{f(tx)-f(t)}{A(t)}-\log x\\&=\frac{a(tx)/a(t)-1}{A(t)/a(t)}+\frac{\int_x^1\frac{a(ts)/a(t)-1}{s}\,ds}{A(t)/a(t)}+(\frac{1}{A(t)/a(t)}-1)\log x,\end{aligned}$$

which gives (2.9) by using (2.7) and (2.10). □

- *Inequality for second order regular variation* (Lemma 5.2 of Draisma et al. [30]). If (2.5) holds, then for any $\epsilon>0$ and $\delta>0$, there exists $t_0>0$ such that for all $0<t\le t_0$ and $0<tx\le t_0$

$$\begin{aligned}&|\frac{\frac{(tx)^{1/\alpha}\bar{F}^-(tx)-t^{1/\alpha}\bar{F}^-(t)}{A(t)}-\frac{x^\rho-1}{\rho}}{B(t)}-\frac{1}{\gamma}(\frac{x^{\rho+\gamma}-1}{\rho+\gamma}-\frac{x^\rho-1}{\rho})|\\&\le\epsilon\left\{1+x^\rho+2x^{\gamma+\rho}e^{\delta|\log x|}\right\}.\end{aligned} \tag{2.11}$$

*Proof.* Put $H(t)=t^{1/\alpha}\bar{F}^-(t)$. When $\rho>0$, it follows from de Haan and Stadtmüller [28] that

$$\frac{(tx)^{-\rho}A(tx)-t^{-\rho}A(t)}{t^{-\rho}A(t)B(t)}\to\frac{x^\gamma-1}{\gamma} \tag{2.12}$$

and

$$\frac{H(tx)-A(tx)/\rho-\{H(t)-A(t)/\rho\}}{A(t)B(t)/\rho}\to-\frac{x^{\rho+\gamma}-1}{\rho+\gamma}. \tag{2.13}$$

Applying (2.9) with $f(x)=x^{-\rho}A(x)$ and $f(x)=H(x)-A(x)/\rho$, respectively, (2.11) follows from the following expression

$$\begin{aligned}&\frac{\frac{H(tx)-H(t)}{A(t)}-\frac{x^\rho-1}{\rho}}{B(t)}-\frac{1}{\gamma}\left\{\frac{x^{\rho+\gamma}-1}{\rho+\gamma}-\frac{x^\rho-1}{\rho}\right\}\\&=\frac{H(tx)-A(tx)/\rho-\{H(t)-A(t)/\rho\}}{A(t)B(t)}+\frac{x^{\rho+\gamma}-1}{\rho(\rho+\gamma)}\\&+x^\rho\left\{\frac{(tx)^{-\rho}A(tx)-t^{-\rho}A(t)}{t^{-\rho}A(t)B(t)\rho}-\frac{x^\gamma-1}{\gamma\rho}\right\}.\end{aligned}$$

When $\rho=0$ and $\gamma<0$, it follows from de Haan and Stadtmüller [28] that

$$A(t)\to c_0\in(0,\infty),\qquad\frac{c_0-A(t)}{A(t)B(t)}\to-\frac{1}{\gamma}, \tag{2.14}$$

and

$$\frac{H(tx) - c_0 \log(tx) - \{H(t) - c_0 \log t\}}{A(t)B(t)} \to \frac{1}{\gamma}\frac{x^\gamma - 1}{\gamma} \tag{2.15}$$

as $t \to 0$. Applying (2.9) with $f(x) = H(x) - c_0 \log(x)$, (2.11) follows from (2.14) and the following expression

$$\begin{aligned} &\frac{\frac{H(tx)-H(t)}{A(t)} - \log x}{B(t)} - \frac{1}{\gamma}\left\{\frac{x^\gamma - 1}{\gamma} - \log x\right\} \\ &= \frac{H(tx) - c_0 \log(tx) - \{H(t) - c_0 \log t\}}{A(t)B(t)} - \frac{1}{\gamma}\frac{x^\gamma - 1}{\gamma} \\ &\quad + \left\{\frac{c_0 - A(t)}{A(t)B(t)} + \frac{1}{\gamma}\right\} \log x. \end{aligned}$$

When $\rho = \gamma = 0$, put $h(t) = H(t) - t^{-1}\int_t^1 H(s)\,ds$. Then $H(t) = h(t) + \int_t^1 \frac{h(s)}{s}\,ds$ and it follows from de Haan and Stadtmüller [28] that

$$\lim_{t \to 0} \frac{h(tx) - h(t)}{A(t)B(t)} = \log x \quad \text{for all} \quad x > 0. \tag{2.16}$$

Write

$$\begin{aligned} &\frac{\frac{H(tx)-H(t)}{A(t)} - \log x}{B(t)} - \frac{1}{2}(\log x)^2 \\ &= \left\{\frac{h(tx) - h(t)}{A(t)B(t)} - \log x\right\} - \int_x^1 \frac{1}{s}\left\{\frac{h(ts) - h(t)}{A(t)B(t)} - \log s\right\} ds \\ &\quad - \left\{\frac{h(t) - A(t)}{A(t)B(t)} - 1\right\} \log x. \end{aligned} \tag{2.17}$$

Then it follows from (2.5), (2.16) and (2.17) that

$$\frac{h(t) - A(t)}{A(t)B(t)} \to 1 \quad \text{as} \quad t \to 0. \tag{2.18}$$

Applying (2.9) with $f(x) = h(x)$, (2.11) follows from (2.17) and (2.18). □

Since some statistics in analyzing extremes are constructed in terms of logarithms of data, it becomes convenient to express (2.2) and (2.4) as

$$\lim_{t \to 0} \left\{\log \bar{F}^-(tx) - \log \bar{F}^-(t)\right\} = -\frac{1}{\alpha} \log x \quad \text{for all} \quad x > 0, \tag{2.19}$$

and

$$\lim_{t\to 0}\frac{\log \bar{F}^-(tx)-\log \bar{F}^-(t)+\alpha^{-1}\log x}{A(t)/\left\{t^{1/\alpha}\bar{F}^-(t)\right\}}=\frac{x^\rho-1}{\rho}\quad \text{for all}\quad x>0, \tag{2.20}$$

respectively. However, when $\rho=\gamma$, (2.5) may not imply a second order regular variation condition for $\log \bar{F}^-(t)$; see Theorem A of Draisma et al. [30]. The following theorem is slightly different from Theorem A of Draisma et al. [30], and shows that (2.5) does imply a corresponding result for $\log \bar{F}^-(t)$ under an additional condition. This theorem is useful in the study of bias corrected tail index estimation.

**Theorem 2.1.** *Suppose (2.5) holds with $\rho>0$ and $\gamma\geq 0$. Further assume*

$$\lim_{t\to 0}\frac{A(t)}{t^{1/\alpha}\bar{F}^-(t)B(t)}=l_0\in[-\infty,\infty].$$

*Then*

$$\lim_{t\to 0}\frac{\frac{\log \bar{F}^-(tx)-\log \bar{F}^-(t)+\alpha^{-1}\log x}{A(t)/(t^{1/\alpha}\bar{F}^-(t))}-\frac{x^\rho-1}{\rho}}{B(t)+A(t)/(t^{1/\alpha}\bar{F}^-(t))} \\ =\frac{H_{\rho,\gamma}(x)}{1+l_0}-\frac{l_0}{2(1+l_0)}(\frac{x^\rho-1}{\rho})^2, \tag{2.21}$$

*where $H_{\rho,\gamma}(x)$ is given in (2.5). Moreover, for any $\epsilon>0$ and $\delta>0$ there exists $t_0>0$ such that for all $0<t\leq t_0$ and $0<tx\leq t_0$*

$$|\frac{\frac{\log \bar{F}^-(tx)-\log \bar{F}^-(t)+\alpha^{-1}\log x}{A(t)/(t^{1/\alpha}\bar{F}^-(t))}-\frac{x^\rho-1}{\rho}}{B(t)+A(t)/(t^{1/\alpha}\bar{F}^-(t))}-\frac{H_{\rho,\gamma}(x)}{1+l_0}+\frac{l_0}{2(1+l_0)}(\frac{x^\rho-1}{\rho})^2| \\ \leq\epsilon\left\{1+x^\rho+2x^{\rho+\gamma}e^{\delta|\log x|}\right\}. \tag{2.22}$$

*Proof.* Put $f(t)=t^{1/\alpha}\bar{F}^-(t)$. Then (2.5) implies that

$$\lim_{t\to 0}\frac{f(tx)-f(t)}{A(t)}=\frac{x^\rho-1}{\rho},$$

which implies that $\lim_{t\to 0}f(t)=c_0\in(0,\infty)$ since $\rho>0$. By (2.5) we have

$$\frac{f(tx)}{f(t)}-1=\frac{A(t)}{f(t)}\frac{x^\rho-1}{\rho}+\frac{A(t)B(t)}{f(t)}H_{\rho,\gamma}(x)(1+o(1)),$$

which implies that

$$\log\frac{f(tx)}{f(t)} = \frac{A(t)}{f(t)}\frac{x^\rho-1}{\rho} + \frac{A(t)B(t)}{f(t)}H_{\rho,\gamma}(x)(1+o(1)) \\ -\frac{1}{2}\frac{A^2(t)}{f^2(t)}(\frac{x^\rho-1}{\rho})^2(1+o(1)),$$

i.e., (2.21) holds.

When $|l_0|<\infty$, we have $\gamma=\rho$ and

$$\frac{H_{\rho,\gamma}(x)}{1+l_0} - \frac{1}{2}\frac{l_0}{1+l_0}(\frac{x^\rho-1}{\rho})^2 = \frac{H_{\rho,\gamma}(x)}{1+l_0} - \frac{l_0H_{\rho,\gamma}(x)}{1+l_0} = \frac{1-l_0}{1+l_0}H_{\rho,\gamma}.$$

Hence, when $|l_0|<\infty$, (2.22) follows from applying (2.11) to the function $\log f(t)$. Other cases can be shown similarly by noticing that

$$\{A(t)/f(t)+B(t)\}\in RV_\rho^0 \quad \text{in case of} \quad |l_0|=\infty. \qquad \square$$

Other important techniques in studying extremes are the following empirical process, quantile process, tail empirical process and tail quantile process.

Let $U_1,\cdots,U_n$ be independent and identically distributed random variables with uniform distribution on $(0,1)$, and let $U_{n,1}\le\cdots\le U_{n,n}$ denote the order statistics of $U_1,\cdots,U_n$. Therefore,

$$G_n(u)=\frac{1}{n}\sum_{i=1}^n I(U_i\le u) \quad \text{and} \quad \alpha_n(u)=\sqrt{n}\{G_n(u)-u\}$$

are called the **empirical distribution function** and **empirical process**, respectively. By the Chibisov–O'Reilly theorem,

$$\sup_{0<u<1}\frac{|\alpha_n(u)|}{u^{1/4}}=O_p(1) \quad \text{and} \quad \sup_{0<u<1}\frac{|\alpha_n\big((1-u)-\big)|}{u^{1/4}}=O_p(1), \tag{2.23}$$

where $\alpha_n(u-)$ denotes the left-hand limit of $\alpha_n(u)$. It also follows from Shorack and Wellner [98], Page 404 that

$$\sup_{U_{n,1}\le u\le 1}\frac{u}{G_n(u)}=O_p(1), \qquad \sup_{1-U_{n,n}\le u\le 1}\frac{u}{1-G_n\big((1-u)-\big)}=O_p(1), \tag{2.24}$$

$$\sup_{0<u<1}\frac{G_n(u)}{u}=O_p(1) \quad \text{and} \quad \sup_{0<u<1}\frac{1-G_n(1-u)}{u}=O_p(1). \tag{2.25}$$

By Csörgő et al. [25], there exists a sequence of Brownian bridges $\{B_n(u)\}$ such that for any $\nu \in [0, 1/4)$

$$\sup_{U_{n,1}\le u\le U_{n,n}} \frac{n^{\nu}|\alpha_n(u)-B_n(u)|}{u^{1/2-\nu}(1-u)^{1/2-\nu}} = O_p(1). \tag{2.26}$$

Put $Q_n(s) = U_{n,k}$ if $\frac{k-1}{n} < s \le \frac{k}{n}$ for $k = 1, \cdots, n$, and $Q_n(0) = U_{n,1}$, which is called the **quantile function**. Further $\beta_n(s) = \sqrt{n}\{Q_n(s) - s\}$ is called the **quantile process**. Again it follows from Csörgő et al. [25] that for any $\nu \in [0, 1/2)$ and $\lambda > 0$

$$\sup_{\lambda/n\le s\le 1-\lambda/n} \frac{n^{\nu}|\beta_n(s)+B_n(s)|}{s^{1/2-\nu}(1-s)^{1/2-\nu}} = O_p(1), \tag{2.27}$$

where $B_n(s)$ is given in (2.26).

**Theorem 2.2.** *Define the* ***tail empirical process*** *as*

$$\alpha_{n,k}(u) = \sqrt{k}\left\{\frac{1}{k}\sum_{i=1}^{n} I(U_i \le \frac{k}{n}u) - u\right\} \quad \textit{for} \quad u > 0,$$

*where k satisfies*

$$k = k(n) \to \infty \quad \textit{and} \quad k/n \to 0 \quad \textit{as} \quad n \to \infty. \tag{2.28}$$

*Then for any* $\nu \in [0, 1/2)$

$$\sup_{0<u\le 1} \frac{|\alpha_{n,k}(u) - W_n(u)|}{u^{\nu}} \xrightarrow{p} 0 \quad \textit{as} \quad n \to \infty, \tag{2.29}$$

*where* $W_n(u) = \sqrt{\frac{n}{k}} B_n(\frac{k}{n}u)$ *with* $B_n$ *given in (2.26).*

*Proof.* Note that

$$\begin{aligned}\frac{\alpha_{n,k}(u) - W_n(u)}{u^{\nu}} &= \sqrt{\frac{n}{k}}\frac{\alpha_n(\frac{k}{n}u) - B_n(\frac{k}{n}u)}{u^{\nu}} \\ &= k^{-1/2+\nu}\frac{n^{1/2-\nu}\left\{\alpha_n(\frac{k}{n}u) - B_n(\frac{k}{n}u)\right\}}{(\frac{k}{n}u)^{\nu}}.\end{aligned}$$

Hence (2.29) follows from (2.26) and (2.28) when $\nu$ is close to $1/2$ and $u \ge nU_{n,1}/k$. When $u < nU_{n,1}/k$, we have

$$\frac{|\alpha_{n,k}(u) - W_n(u)|}{u^{\delta}} \le \sqrt{k}u^{1-\delta} + u^{1/2-\delta}|\sqrt{\frac{n}{ku}}B_n(\frac{k}{n}u)|$$

$$\leq k^{-1/2+\delta}(nU_{n,1})^{1-\delta} + u^{1/2-\delta}|\sqrt{\frac{n}{ku}}B_n(\frac{k}{n}u)|$$

$$= o_p(1).$$

Therefore (2.29) holds for $\nu$ near 1/2, which implies that it holds for any $\nu \in [0, 1/2)$. □

**Theorem 2.3.** *Define the* ***tail quantile process*** *as*

$$\beta_{n,k}(s) = \sqrt{k}\left\{\frac{n}{k}U_{n,[ks]} - s\right\} \quad \textit{for} \quad s > 0,$$

*where $[x]$ denotes the smallest positive integer larger than or equal to $x$, and $k$ satisfies (2.28). Then, for any $\nu \in [0, 1/2)$*

$$\sup_{1 \leq sk \leq n-1} \frac{|\beta_{n,k}(s) + W_n(s)|}{s^\nu} \xrightarrow{p} 0 \quad \textit{as} \quad n \to \infty, \tag{2.30}$$

*where $W_n(s)$ is given in (2.29).*

*Proof.* Write

$$\frac{\beta_{n,k}(s) + W_n(s)}{s^\nu} = k^{-1/2+\nu}\frac{n^{1/2-\nu}\left\{\beta_n(\frac{k}{n}s) + B_n(\frac{k}{n}s)\right\}}{(\frac{k}{n}s)^\nu}.$$

Then (2.30) follows from (2.27). □

**Remark 2.1.** Note that $\{W_n(s)\}$ in (2.29) and (2.30) can be replaced by a sequence of standard Wiener processes.

## 2.2 TAIL INDEX ESTIMATION

Throughout this section, we assume $X, X_1, \cdots, X_n$ are independent and identically distributed random variables with distribution function $F(x)$ satisfying (2.1). Let $X_{n,1} \leq \cdots \leq X_{n,n}$ denote the order statistics of $X_1, \cdots, X_n$, and let $F_n(x) = \frac{1}{n}\sum_{i=1}^n I(X_i \leq x)$ denote the empirical distribution function. Assume $U_1, \cdots, U_n$ are independent and identically distributed random variables with uniform distribution on (0, 1) and let $U_{n,1} \leq \cdots \leq U_{n,n}$ denote the order statistics of $U_1, \cdots, U_n$. It follows from Lemma 1.1 that $\bar{F}^-(U_1), \cdots, \bar{F}^-(U_n)$ are also independent random variables with the distribution function $F(x)$, and

$$\left(F^{-}(U_{n,n}),\cdots,F^{-}(U_{n,1})\right)^T,\quad \left(\bar{F}^{-}(U_{n,1}),\cdots,\bar{F}^{-}(U_{n,n})\right)^T$$
$$\text{and}\quad \left(X_{n,n},\cdots,X_{n,1}\right)^T$$

have the same joint distribution. For convenience, we often write that

$$\left\{X_{n,n-i+1}=\bar{F}^{-}(U_{n,i}):i=1,\cdots,n\right\}$$
$$\text{or}\quad \left\{X_{n,n-i+1}=F^{-}(U_{n,n-i+1}):i=1,\cdots,n\right\}.$$

Although many estimators of the tail index $\alpha$ in (2.1) have been proposed in the literature, we will use the well-known Hill estimator (Hill [56]) to address some important issues and solutions in estimating the tail index $\alpha$ such as optimal choice of the sample fraction $k$, bias corrected estimation, high quantile estimation, interval estimation, and goodness-of-fit tests.

### 2.2.1 Hill Estimator

Using (2.2) and its Potter's bound in (2.6), we have

$$\lim_{t\to 0}\int_0^1 \log\frac{\bar{F}^{-}(tx)}{\bar{F}^{-}(t)}\,dx=-\frac{1}{\alpha}\int_0^1\log x\,dx=\frac{1}{\alpha},$$

which motivates us to estimate $\alpha$ by

$$\hat{\alpha}(k)=\left\{\int_0^1\log\frac{\bar{F}_n^{-}(\frac{k}{n}x)}{\bar{F}_n^{-}(\frac{k}{n})}\,dx\right\}^{-1}=\left\{\frac{1}{k}\sum_{i=1}^{k}\log\frac{X_{n,n-i+1}}{X_{n,n-k}}\right\}^{-1},\tag{2.31}$$

where $k$ satisfies (2.28). This is the so-called Hill estimator in the literature (see Hill [56]). Another useful way to derive the above Hill estimator $\hat{\alpha}(k)$ is via maximizing a censored likelihood function as follows.

Define $\delta_i=I(X_i>T)$ for a high threshold $T$ and temporarily assume the conditional distribution of $X_i$ given $\delta_i=1$ is $1-cx^{-\alpha}$. Here $I(A)$ denotes the indicator function of the set $A$. Then a censored likelihood function for $\left\{(X_i,\delta_i)^T\right\}_{i=1}^n$ can be written as

$$\prod_{i=1}^{n}\left\{c\alpha X_i^{-\alpha-1}\right\}^{\delta_i}\left\{1-cT^{-\alpha}\right\}^{1-\delta_i},\tag{2.32}$$

which is maximized at

$$c=\frac{\sum_{i=1}^n\delta_i}{n}T^{\alpha}\quad\text{and}\quad \alpha=\frac{\sum_{i=1}^n\delta_i}{\sum_{i=1}^n\delta_i\log(X_i/T)}.\tag{2.33}$$

Therefore the Hill estimator $\hat{\alpha}(k)$ in (2.31) is obtained by taking $T = X_{n,n-k}$ in the second equation of (2.33).

#### 2.2.1.1 Asymptotic Properties

The asymptotic properties of the Hill estimator $\hat{\alpha}(k)$ are given below.

**Theorem 2.4.** *(i) Consistency. Under conditions (2.1) and (2.28), we have*

$$\hat{\alpha}(k) \xrightarrow{p} \alpha \quad as \quad n \to \infty.$$

*(ii) Asymptotic normality. Under conditions (2.1), (2.4), (2.28) and*

$$\sqrt{k}\frac{A(k/n)}{(k/n)^{1/\alpha}\bar{F}^{-}(k/n)} \to \lambda \in \mathbb{R} \quad as \quad n \to \infty,$$

*we have*

$$\sqrt{k}\left\{\hat{\alpha}(k) - \alpha\right\} \xrightarrow{d} N(\frac{\lambda\alpha^2}{1+\rho}, \alpha^2) \quad as \quad n \to \infty.$$

*Proof.* (i) Write $\left\{X_{n,n-i+1}\right\}_{i=1}^{n} = \left\{\bar{F}^{-}(U_{n,i})\right\}_{i=1}^{n}$, which implies that

$$\frac{1}{k}\sum_{i=1}^{k}\log\frac{X_{n,n-i+1}}{X_{n,n-k}} = \frac{1}{k}\sum_{i=1}^{k}\log\frac{\bar{F}^{-}(U_{n,i})}{\bar{F}^{-}(U_{n,k+1})}. \tag{2.34}$$

By (2.6), for any $0 < \epsilon < 1$ and $0 < \delta < 1/2$, there is a $t_0 > 0$ such that, for any $i = 1, \cdots, k$ and $U_{n,k+1} \le t_0$

$$\begin{aligned}
&\log(1-\epsilon) - \frac{1}{\alpha}\log\frac{U_{n,i}}{U_{n,k+1}} + (\frac{U_{n,i}}{U_{n,k+1}})^{\delta} \\
&\le \log\bar{F}^{-}(U_{n,i}) - \log\bar{F}^{-}(U_{n,k+1}) \\
&\le \log(1+\epsilon) - \frac{1}{\alpha}\log\frac{U_{n,i}}{U_{n,k+1}} + (\frac{U_{n,i}}{U_{n,k+1}})^{-\delta}.
\end{aligned} \tag{2.35}$$

Put $G_{n,k}(s) = \frac{1}{k}\sum_{i=1}^{n} I(U_i \le \frac{k}{n}s)$. It follows from (2.29) and (2.30) that

$$\frac{n}{k}U_{n,k} \xrightarrow{p} 1, \quad \frac{n}{k}U_{n,k+1} \xrightarrow{p} 1 \quad \text{and} \quad \sup_{0<s\le 1}\frac{|G_{n,k}(s) - s|}{s^{\delta}} \xrightarrow{p} 0 \quad \text{as} \quad n \to \infty,$$

which imply that

$$\begin{aligned}
&\frac{1}{k}\sum_{i=1}^{k}(\frac{U_{n,i}}{U_{n,k+1}})^{-\delta}=\int_0^{\frac{n}{k}U_{n,k}}(\frac{\frac{k}{n}s}{U_{n,k+1}})^{-\delta}\,dG_{n,k}(s)\\
=&\ (\frac{n}{k}U_{n,k+1})^{\delta}\left\{(\frac{n}{k}U_{n,k})^{-\delta}G_{n,k}(\frac{n}{k}U_{n,k})+\delta\int_0^{\frac{n}{k}U_{n,k}}G_{n,k}(s)s^{-\delta-1}\,ds\right\} \qquad (2.36)\\
\stackrel{p}{\to}&\ 1+\delta\textstyle\int_0^1 s^{-\delta}\,ds=\dfrac{1}{1-\delta}.
\end{aligned}$$

Similarly we can show that as $n\to\infty$

$$\begin{cases}\dfrac{1}{k}\displaystyle\sum_{i=1}^{k}\log\frac{U_{n,i}}{U_{n,k+1}}=(1+o_p(1))\int_0^1\log s\,ds=-1+o_p(1),\\[2ex] \dfrac{1}{k}\displaystyle\sum_{i=1}^{k}(\frac{U_{n,i}}{U_{n,k+1}})^{\delta}=(1+o_p(1))\int_0^1 s^{\delta}\,ds=\frac{1}{1+\delta}+o_p(1).\end{cases} \qquad (2.37)$$

Hence by using (2.34)–(2.37) we have

$$|\frac{1}{k}\sum_{i=1}^{k}\log\frac{X_{n,n-i+1}}{X_{n,n-k}}-\frac{1}{\alpha}|\le\frac{\delta}{1-\delta}-\log(1-\epsilon)+o_p(1).$$

Then by letting both $\delta$ and $\epsilon$ tend to zero, we obtain

$$\frac{1}{k}\sum_{i=1}^{k}\log\frac{X_{n,n-i+1}}{X_{n,n-k}}\stackrel{p}{\to}\frac{1}{\alpha}\quad\text{as}\quad n\to\infty,$$

i.e., the claimed consistency holds.

(ii) Note that (2.4) implies (2.20), i.e.,

$$\lim_{t\to 0}\frac{\log\left\{(tx)^{1/\alpha}\bar{F}^{-}(tx)\right\}-\log\left\{t^{1/\alpha}\bar{F}^{-}(t)\right\}}{A(t)/\left\{t^{1/\alpha}\bar{F}^{-}(t)\right\}}=\frac{x^{\rho}-1}{\rho}\quad\text{for all}\quad x>0. \qquad (2.38)$$

Applying (2.9) with $f(x)=\log\left\{x^{1/\alpha}\bar{F}^{-}(x)\right\}$ and using (2.34) and (2.30), we can show that

$$\begin{aligned}
&\sqrt{k}\left\{\frac{1}{k}\sum_{i=1}^{k}\log\frac{X_{n,n-i+1}}{X_{n,n-k}}-\frac{1}{\alpha}\right\}=\frac{\sqrt{k}}{\alpha}\left\{-\frac{1}{k}\sum_{i=1}^{k}\log\frac{U_{n,i}}{U_{n,k+1}}-1\right\}\\
&+\sqrt{k}\frac{A(U_{n,k+1})}{U_{n,k+1}^{1/\alpha}\bar{F}^{-}(U_{n,k+1})}\frac{1}{k}\sum_{i=1}^{k}\frac{(U_{n,i}/U_{n,k+1})^{\rho}-1}{\rho}(1+o_p(1)),
\end{aligned}$$

$$\begin{aligned}
&\frac{\sqrt{k}}{\alpha}\left\{-\frac{1}{k}\sum_{i=1}^{k}\log\frac{U_{n,i}}{U_{n,k+1}}-1\right\}\\
&=-\frac{\sqrt{k}}{\alpha}\frac{1}{k}\sum_{i=1}^{k}\log\frac{\frac{n}{i}U_{n,i}}{\frac{n}{k+1}U_{n,k+1}}+\frac{\sqrt{k}}{\alpha}\left\{\frac{1}{k}\sum_{i=1}^{k}\log\frac{k+1}{i}-1\right\}\\
&=-\frac{\sqrt{k}}{\alpha}\frac{1}{k}\sum_{i=1}^{k}\left\{\frac{\frac{n}{i}U_{n,i}}{\frac{n}{k+1}U_{n,k+1}}-1\right\}+o_p(1)\\
&=-\alpha^{-1}\left\{\int_0^1\frac{\beta_{n,k}(s)}{s}\,ds-\beta_{n,k}(1)\right\}+o_p(1)\\
&=\alpha^{-1}\int_0^1\left\{\frac{W_n(s)}{s}-W_n(1)\right\}ds+o_p(1)\\
&\xrightarrow{d}N(0,\tau^2)
\end{aligned}$$

with

$$\begin{aligned}
\tau^2&=\alpha^{-2}E\int_0^1\int_0^1\left\{\frac{W_1(s)W_1(t)}{st}-\frac{W_1(t)W_1(1)}{t}-\frac{W_1(s)W_1(1)}{s}+W_1^2(1)\right\}dsdt\\
&=\alpha^{-2}\int_0^1\int_0^1\left\{\frac{\min(s,t)}{st}-1\right\}dsdt\\
&=\alpha^{-2},
\end{aligned}$$

and

$$\frac{1}{k}\sum_{i=1}^{k}\frac{(U_{n,i}/U_{n,k+1})^{\rho}-1}{\rho}\xrightarrow{p}\int_0^1\frac{s^{\rho}-1}{\rho}\,ds=-\frac{1}{1+\rho},$$

which imply that

$$\sqrt{k}\left\{\frac{1}{k}\sum_{i=1}^{k}\log\frac{X_{n,n-i+1}}{X_{n,n-k}}-\alpha^{-1}\right\}\xrightarrow{d}N\left(-\frac{\lambda}{1+\rho},\alpha^{-2}\right)\quad\text{as}\quad n\to\infty,$$

i.e., the claimed asymptotic normality holds. □

#### 2.2.1.2 Optimal Choice of $k$

It follows from Theorem 2.4 that a larger $k$ leads to a smaller variance and a bigger bias for the Hill estimator $\hat{\alpha}(k)$. When (2.4) holds with $A(t)/\left\{t^{1/\alpha}\bar{F}^{-}(t)\right\}=dt^{\rho}$ and some $\rho>0$, a theoretical optimal $k$ in $\hat{\alpha}(k)$ can be chosen to balance the bias and variance in the sense of minimizing the asymptotic mean squared error of $\hat{\alpha}(k)$. That is,

$$k_{opt} = \arg\min_{k} \left\{ \frac{\alpha^4}{(1+\rho)^2} d^2 (\frac{k}{n})^{2\rho} + \frac{\alpha^2}{k} \right\} = \left\{ \frac{(1+\rho)^2}{2\rho d^2 \alpha^2} \right\}^{\frac{1}{1+2\rho}} n^{\frac{2\rho}{1+2\rho}}, \qquad (2.39)$$

which depends on the tail index $\alpha$ and the second order regular variation parameters $\rho$ and $d$.

#### 2.2.1.3 Data Driven Methods for Choosing $k$

Since the asymptotic distribution of $\hat{\alpha}(k)$ in Theorem 2.4 depends on a second order regular variation condition, it is extremely challenging to choose $k$ in practice. In general one should plot the tail index estimator against various $k$'s to identify a relatively stable region with respect to $k$. If the sample size $n$ is large, some data-driven procedures for choosing $k$ are available in the literature.

**Method 1) direct estimation.** A simple way for choosing $k$ is to estimate the optimal $k_{opt}$ in (2.39) directly, which requires consistent estimators for the nuisance parameters $d$ and $\rho$.

*1a) A class of estimators for $\rho$ in Gomes et al. [44].* For $\delta > 0$, define

$$M_n^{(\delta)}(k) = \frac{1}{k} \sum_{i=1}^{k} \left\{ \log \frac{X_{n,n-i+1}}{X_{n,n-k}} \right\}^{\delta},$$

$$Q_n^{(\delta)}(k) = \frac{M_n^{(\delta)}(k) - \Gamma(\delta+1) \left\{ M_n^{(1)}(k) \right\}^{\delta}}{M_n^{(2)}(k) - 2 \left\{ M_n^{(1)}(k) \right\}^2},$$

$$S_n^{(\delta)}(k) = \frac{\delta(\delta+1)^2 \Gamma^2(\delta)}{4\Gamma(2\delta)} \frac{Q_n^{(2\delta)}(k)}{\left\{ Q_n^{(\delta+1)}(k) \right\}^2}$$

and

$$s_\delta(\rho) = \frac{\rho^2 \left\{ 1 - (1+\rho)^{2\delta} + 2\delta\rho(1+\rho)^{2\delta-1} \right\}}{\left\{ 1 - (1+\rho)^{\delta+1} + (\delta+1)\rho(1+\rho)^{\delta} \right\}^2},$$

where $\Gamma(a) = \int_0^\infty x^{a-1} e^{-x} dx$ is the gamma function. Then a class of estimators for $\rho$ is defined as

$$\hat{\rho}^{(\delta)}(k) = s_\delta^{-}(S_n^{(\delta)}(k)) \quad \text{for any} \quad \delta > 0 \quad \text{except} \quad \frac{1}{2} \quad \text{and} \quad 1.$$

**Lemma 2.1.** *The above function $s_\delta(\rho)$ is an increasing function of $\rho > 0$ when $\delta > 1/2$ and $\delta \neq 1$, and a decreasing function of $\rho > 0$ when $\delta \in (0, 1/2)$. Moreover*

$$\lim_{\rho\to 0} s_\delta(\rho) = \frac{4(2\delta-1)}{\delta(\delta+1)^2} \quad \textit{and} \quad \lim_{\rho\to\infty} s_\delta(\rho) = \frac{2\delta-1}{\delta^2}. \tag{2.40}$$

*Proof.* First it is easy to verify (2.40). Second for proving the monotonicity of $s_\delta(\rho)$, we only consider the case of $\delta > 1/2$ and $\delta \neq 1$ since the proof for the case of $\delta \in (0, 1/2)$ is the same.

When $\delta > 1/2$, we have $\frac{4(2\delta-1)}{\delta(\delta+1)^2} < \frac{2\delta-1}{\delta^2}$. Put $x = 1+\rho$ and let $s \in (\frac{4(2\delta-1)}{\delta(\delta+1)^2}, \frac{2\delta-1}{\delta^2})$ be any given number. Then the monotonicity of $s_\delta(\rho)$ is equivalent to showing that there is a unique solution to

$$\begin{aligned} f_1(x) := &(x-1)^2\left\{1 - x^{2\delta} + 2\delta(x-1)x^{2\delta-1}\right\} \\ &- s\left\{1 - x^{\delta+1} + (\delta+1)(x-1)x^\delta\right\}^2 \\ = &\,0. \end{aligned}$$

Define

$$\begin{aligned} f_2(x) := \frac{f_1'(x)}{2(x-1)} &= 1 - x^{2\delta} + 2\delta(x-1)x^{2\delta-1} + \delta(2\delta-1)(x-1)^2x^{2\delta-2} \\ &\quad - s\delta(\delta+1)x^{\delta-1}\left\{1 - x^{\delta+1} + (\delta+1)(x-1)x^\delta\right\}, \\ f_3(x) := \frac{f_2'(x)}{\delta x^{\delta-2}} &= 4(2\delta-1)(x-1)x^\delta + 2(2\delta-1)(\delta-1)(x-1)^2x^{\delta-1} \\ &\quad - s(\delta+1)(\delta-1)\left\{1 - x^{\delta+1} + (\delta+1)(x-1)x^\delta\right\} \\ &\quad - s\delta(\delta+1)^2(x-1)x^\delta, \\ f_4(x) := \frac{f_3'(x)}{x^{\delta-2}} &= 2(2\delta-1)(\delta-1)^2 - 4\delta^2(2\delta-1)x + 2(2\delta-1)(\delta+1)^2x^2 \\ &\quad - s\delta(\delta+1)^2\left\{2\delta x^2 - (2\delta-1)x\right\}. \end{aligned}$$

Then

$$f_4'(x) = -4\delta^2(2\delta-1) + s\delta(\delta+1)^2(2\delta-1) + 4(\delta+1)^2x\left\{2\delta - 1 - s\delta^2\right\}.$$

Since $s < \frac{2\delta-1}{\delta^2}$, we know $f_4'(x)$ is an increasing function of $x$. Now

$$f_4'(1) = (2\delta+1)\delta(\delta+1)^2\left\{\frac{4(2\delta-1)}{\delta(\delta+1)^2} - s\right\} < 0 \quad \text{and} \quad f_4'(\infty) = \infty,$$

there exists an $x_0 \in (1, \infty)$ such that

$$f_4'(x) < 0 \quad \text{for} \quad x \in (1, x_0) \quad \text{and} \quad f_4'(x) > 0 \quad \text{for} \quad x \in (x_0, \infty).$$

By noting that

$$f_4(1) = \delta(\delta+1)^2 \left\{ \frac{4(2\delta-1)}{\delta(\delta+1)^2} - s \right\} < 0 \quad \text{and} \quad f_4(\infty) = \infty,$$

there exists an $x_1 \in (x_0, \infty)$ such that

$$f_3'(x) < 0 \quad \text{for} \quad x \in (1, x_1) \quad \text{and} \quad f_3'(x) > 0 \quad \text{for} \quad x \in (x_1, \infty).$$

Use the facts that

$$\lim_{x\to 1} \frac{f_3(x)}{x-1} = \lim_{x\to 1} f_3'(x) = f_4(1) < 0 \quad \text{and} \quad \lim_{x\to\infty} f_3(x) = \infty$$

since

$$\lim_{x\to\infty} \frac{f_3(x)}{x^{\delta+1}} = 2(\delta+1)\delta^2 \left\{ \frac{2\delta-1}{\delta^2} - s \right\} > 0,$$

we conclude that there exists an $x_2 \in (x_1, \infty)$ such that

$$f_2'(x) < 0 \quad \text{for} \quad x \in (1, x_2) \quad \text{and} \quad f_2'(x) > 0 \quad \text{for} \quad x \in (x_2, \infty).$$

Again, by noting that

$$\lim_{x\to\infty} \frac{f_2(x)}{x^{2\delta}} = \delta(\delta+1) \left\{ \frac{2\delta-1}{\delta^2} - s \right\} > 0,$$

we have $\lim_{x\to\infty} f_2(x) = \infty$ and

$$\lim_{x\to 1} \frac{f_2(x)}{(x-1)^2} = \lim_{x\to 1} \frac{f_2'(x)}{2(x-1)} = \lim_{x\to 1} \frac{\delta f_3(x)}{2(x-1)} = \frac{\delta}{2} \lim_{x\to 1} f_3'(x) = \frac{\delta}{2} f_4(1) < 0,$$

which imply that there exists an $x_3 \in (x_2, \infty)$ such that

$$f_1'(x) < 0 \quad \text{for} \quad x \in (1, x_3) \quad \text{and} \quad f_1'(x) > 0 \quad \text{for} \quad x \in (x_3, \infty).$$

Since

$$\lim_{x\to\infty} \frac{f_1(x)}{x^{2\delta+2}} = \delta^2 \left\{ \frac{2\delta-1}{\delta^2} - s \right\} > 0,$$

we have $\lim_{x\to\infty} f_1(x) = \infty$ and

$$\lim_{x\to 1} \frac{f_1(x)}{(x-1)^4} = \lim_{x\to 1} \frac{f_1'(x)}{4(x-1)^3} = \lim_{x\to 1} \frac{f_2(x)}{2(x-1)^2} = \frac{\delta}{4} f_4(1) < 0,$$

which imply that there exists a unique $x_4 \in (x_3, \infty)$ such that $f_1(x_4) = 0$. Hence the lemma follows. □

**Theorem 2.5.** *Under conditions (2.4) with some $\rho > 0$, (2.28) and*

$$\sqrt{k}|A(k/n)| \to \infty \quad as \quad n \to \infty, \tag{2.41}$$

*we have $\hat{\rho}^{(\delta)}(k) \xrightarrow{p} \rho$ for any $\delta > 0$ except $\frac{1}{2}$ and 1 as $n \to \infty$.*

*Proof.* With $\rho > 0$, we have $\lim_{t\to 0} t^{1/\alpha}\bar{F}^{-}(t) = c \in (0, \infty)$. Since (2.4) implies (2.38), applying (2.9) with $f(x) = \log\left\{x^{1/\alpha}\bar{F}^{-}(x)\right\}$ and using (2.30), we have

$$\begin{aligned}
M_n^{(\delta)}(k) &= \frac{1}{k}\sum_{i=1}^{k}\left\{-\frac{1}{\alpha}\log\frac{U_{n,i}}{U_{n,k+1}}\right\}^{\delta} + \frac{1}{k}\sum_{i=1}^{k}\delta\left\{-\frac{1}{\alpha}\log\frac{U_{n,i}}{U_{n,k+1}}\right\}^{\delta-1} \\
&\quad \times \frac{(U_{n,i}/U_{n,k+1})^{\rho}-1}{\rho}\frac{A(k/n)}{(k/n)^{1/\alpha}\bar{F}^{-}(k/n)}\left\{1+o_p(1)\right\} \\
&= \left\{1+O_p(1/\sqrt{k})\right\}\int_0^1\left\{-\frac{1}{\alpha}\log s\right\}^{\delta} ds \\
&\quad + \left\{1+o_p(1)\right\}\frac{\delta A(k/n)}{c}\int_0^1\left\{-\frac{1}{\alpha}\log s\right\}^{\delta-1}\frac{s^{\rho}-1}{\rho} ds \\
&= \alpha^{-\delta}\Gamma(\delta+1) + O_p(1/\sqrt{k}) \\
&\quad + \left\{1+o_p(1)\right\}\frac{A(k/n)}{c}\frac{\delta\alpha^{-\delta+1}\Gamma(\delta)\left\{(1+\rho)^{-\delta}-1\right\}}{\rho},
\end{aligned} \tag{2.42}$$

which implies that

$$\begin{aligned}
&M_n^{(\delta)}(k) - \Gamma(\delta+1)\left\{M_n^{(1)}(k)\right\}^{\delta} \\
&= \delta\alpha^{-\delta+1}\Gamma(\delta)\frac{(1+\rho)^{-\delta}-1-\delta(1+\rho)^{-1}+\delta}{\rho}\frac{A(k/n)}{c}\left\{1+o_p(1)\right\},
\end{aligned}$$

i.e.,

$$Q_n^{(\delta)}(k) = \frac{\delta\alpha^{-\delta+2}\Gamma(\delta)\left\{(1+\rho)^{-\delta}-1-\delta(1+\rho)^{-1}+\delta\right\}}{2\rho^2(1+\rho)^{-2}} + o_p(1),$$

i.e.,

$$S_n^{(\delta)}(k) = s_{\delta}(\rho) + o_p(1).$$

Hence the consistency follows from the monotonicity of $s_{\delta}(\rho)$ showed in Lemma 2.1. □

**Remark 2.2.** In practice, one may simply choose $\delta = 2$, which gives

$$s_2(\rho) = (3\rho^2 + 8\rho + 6)/(3 + 2\rho)^2 \quad \text{and} \quad s_2^{-}(s) = \frac{2(3s-2) + \sqrt{3s-2}}{3-4s}$$

for $\rho > 0$ and $2/3 < s < 3/4$.

*1b) A class of estimators for d in Caeiro and Gomes [11].* For $\tau \neq 0$, define

$$\hat{d}_\tau(k) = -\frac{2(2+\tilde{\rho}_n)}{\tau\tilde{\rho}_n\tilde{\alpha}_n}(\frac{n}{k})^{\tilde{\rho}_n}\frac{\{(M_n^{(1)}(k))^\tau - (M_n^{(2)}(k)/2)^{\tau/2}\}^2}{(M_n^{(2)}(k)/2)^\tau - (M_n^{(4)}(k)/24)^{\tau/2}},$$

where $\tilde{\alpha}_n$ and $\tilde{\rho}_n$ are consistent estimators for $\alpha$ and $\rho$, respectively. In practice, one chooses $\tau \in [-2, -0.5]$ as suggested by Caeiro and Gomes [11].

**Theorem 2.6.** *Under conditions (2.28), (2.4) with $A(t)/\left\{t^{1/\alpha}\bar{F}^-(t)\right\} = dt^\rho$ for some $d \neq 0$, $\rho > 0$, $\sqrt{k}|A(k/n)| \to \infty$, $\tilde{\alpha}_n - \alpha = o_p(1)$ and $(\tilde{\rho}_n - \rho)\log(n/k) = o_p(1)$, we have*

$$\hat{d}_\tau(k) \xrightarrow{p} d \quad \textit{for any} \quad \tau \neq 0 \quad \textit{as} \quad n \to \infty.$$

*Proof.* Put $\bar{A} = \frac{A(k/n)}{c}$, where $c = \lim_{t\to 0} t^{1/\alpha}\bar{F}^-(t) \in (0,\infty)$ implied by $\rho > 0$. Using (2.42), we have

$$\left\{M_n^{(1)}(k)\right\}^\tau = \alpha^{-\tau} + O_p(1/\sqrt{k}) + \tau\alpha^{-\tau+1}\frac{(1+\rho)^{-1}-1}{\rho}\bar{A}\left\{1+o_p(1)\right\},$$

$$\left\{\frac{M_n^{(2)}(k)}{2}\right\}^{\tau/2} = \alpha^{-\tau} + O_p(1/\sqrt{k}) + \tau\alpha^{-\tau+1}\frac{(1+\rho)^{-2}-1}{2\rho}\bar{A}\left\{1+o_p(1)\right\},$$

$$\left\{\frac{M_n^{(2)}(k)}{2}\right\}^{\tau} = \alpha^{-2\tau} + O_p(1/\sqrt{k}) + \tau\alpha^{-2\tau+1}\frac{(1+\rho)^{-2}-1}{\rho}\bar{A}\left\{1+o_p(1)\right\}$$

and

$$\left\{\frac{M_n^{(4)}(k)}{24}\right\}^{\tau/2} = \alpha^{-2\tau} + O_p(1/\sqrt{k}) + \tau\alpha^{-2\tau+1}\frac{(1+\rho)^{-4}-1}{2\rho}\bar{A}\left\{1+o_p(1)\right\},$$

which give

$$\frac{\{(M_n^{(1)}(k))^\tau - (M_n^{(2)}(k)/2)^{\tau/2}\}^2}{(M_n^{(2)}(k)/2)^\tau - (M_n^{(4)}(k)/24)^{\tau/2}} = -\frac{\tau\alpha\rho}{2(2+\rho)^2}\bar{A}\left\{1+o_p(1)\right\}.$$

Hence the theorem follows by noting that $(n/k)^{\tilde{\rho}_n-\rho} - 1 = o_p(1)$. □

**Method 2) Sequential method in Drees and Kaufmann [33].** For some sequence $r_n \to \infty$, put

$$\bar{k}(r_n) = \min\left\{k \in \{2, 3, \cdots, n\} \mid \max_{2\le i\le k} i^{1/2}|\frac{1}{\hat{\alpha}(i)} - \frac{1}{\hat{\alpha}(k)}| > r_n\right\},$$

where $\hat{\alpha}(k)$ is the Hill estimator given in (2.31). Then the following theorem gives a data-driven method for choosing the optimal $k_{opt}$ in (2.39).

**Theorem 2.7.** *Suppose conditions (2.1) and (2.4) hold with*

$$A(t)/\left\{t^{1/\alpha}\bar{F}^{-}(t)\right\} = dt^{\rho} \quad \textit{for some} \quad d \neq 0 \quad \textit{and} \quad \rho > 0.$$

*Further assume* $r_n \to \infty$, $r_n = o(n^{1/2})$, $(\log\log n)^{1/2} = o(r_n)$, $\tilde{\rho}_n - \rho = o_p(1)$, $\tilde{\alpha}_n - \alpha = o_p(1)$ *as* $n \to \infty$. *Then, for any* $\xi \in (0, 1)$ *and* $(\log\log n)^{1/(2\xi)} = o(r_n)$, *we have*

$$\hat{k}_{DK}/k_{opt} \xrightarrow{p} 1 \quad as \quad n \to \infty,$$

*where* $k_{opt}$ *is given in (2.39) and*

$$\hat{k}_{DK} = [(2\tilde{\rho}_n + 1)^{-1/\tilde{\rho}_n}(2\tilde{\alpha}_n^{-2}\tilde{\rho}_n)^{1/(2\tilde{\rho}_n+1)}(\frac{\bar{k}(r_n^{\xi})}{(\bar{k}(r_n))^{\xi}})^{1/(1-\xi)}].$$

*Proof.* See Drees and Kaufmann [33]. □

**Method 3) Hall's bootstrap method.** Another way of choosing the optimal $k_{opt}$ in (2.39) is to employ a bootstrap method. It is known that a subsample bootstrap method is needed in order to catch the bias term of a tail index estimator. The following subsample bootstrap procedure is due to Hall [51]. Put

$$MSE(n, k) = E\big(\hat{\alpha}(k) - \alpha\big)^2.$$

Draw a resample $\left\{X_1^*, \cdots, X_{n_1}^*\right\}$ from $\{X_1, \cdots, X_n\}$ with a smaller sample size $n_1 = o(n)$. Let $X_{n_1,1}^* \le \cdots \le X_{n_1,n_1}^*$ denote the order statistics of $X_1^*, \cdots, X_{n_1}^*$ and put

$$\hat{\alpha}^*(n_1, k_1) = \left\{\frac{1}{k_1}\sum_{i=1}^{k_1}\log\frac{X_{n_1,n_1-i+1}^*}{X_{n_1,n_1-k_1}^*}\right\}^{-1}.$$

Then the bootstrap estimator of $MSE(n_1, k_1)$ is

$$\widehat{MSE}(n_1, k_1) = E\left\{\big(\hat{\alpha}^*(n_1, k_1) - \hat{\alpha}(k)\big)^2|X_1, \cdots, X_n\right\}.$$

Next choose $\hat{k}_1$ to minimize $\widehat{MSE}(n_1, k_1)$. Note that the above conditional expectation will be computed by an average of $(\hat{\alpha}^*(n_1, k_1) - \hat{\alpha}(k))^2$ via a large number of resampling. When the optimal $k_{opt} = cn^\gamma$ for an unknown $c > 0$, but a known $\gamma \in (0, 1)$, Hall [51] proposed to estimate $k_{opt}$ by

$$\hat{k}_H = \hat{k}_1(n/n_1)^\gamma.$$

**Theorem 2.8.** *Suppose conditions (2.1) and (2.4) hold with* $A(t)/\{t^{1/\alpha}\bar{F}^-(t)\} = dt^\rho$ *for some* $d \neq 0$ *and* $\rho > 0$. *Further assume* $\epsilon n^\gamma \leq k \leq \epsilon^{-1} n^\gamma$ *for some* $\epsilon \in (0, 1)$ *and* $\gamma = \frac{2\rho}{1+2\rho}$ *is known. Then*

$$\hat{k}_H/k_{opt} \xrightarrow{p} 1 \quad as \quad n \to \infty,$$

*where* $k_{opt}$ *is given in (2.39).*

*Proof.* See Hall [51]. □

**Method 4) Double bootstrap method.** To get rid of the constraint of known $\gamma$ in the above bootstrap method, Danielsson et al. [26] proposed to minimize $Q(n, k) = E(\frac{1}{2}\hat{\alpha}^2(k) - \hat{M}(k))^2$ instead of $MSE(n, k)$ since

$$\frac{\arg\min_k Q(n, k)}{\arg\min_k MSE(n, k)} \to (1 + \frac{1}{\rho})^{\frac{2}{1+2\rho}},$$

where

$$\hat{M}(k) = \left\{\frac{1}{k}\sum_{i=1}^{k}(\log\frac{X_{n,n-i+1}}{X_{n,n-k}})^2\right\}^{-1}. \tag{2.43}$$

More specifically, draw a resample $\{X_1^*, \cdots, X_{n_1}^*\}$ from $\{X_1, \cdots, X_n\}$ with a smaller sample size $n_1 = O(n^{1-\delta})$ for some $\delta \in (0, 1/2)$. Define the corresponding estimators of $\hat{\alpha}(k)$ and $\hat{M}(k)$ based on the bootstrap sample as $\hat{\alpha}^*(k)$ and $\hat{M}^*(k)$, and choose

$$\hat{k}_1 = \arg\min_{k_1} E\left\{\left(\frac{1}{2}(\hat{\alpha}^*(k_1))^2 - \hat{M}^*(k_1)\right)^2 | X_1, \cdots, X_n\right\}.$$

The above conditional expectation is computed by an average of $\left\{\frac{1}{2}(\hat{\alpha}^*(k_1))^2 - \hat{M}^*(k_1)\right\}^2$ via a large number of resampling. Repeat the above procedure with $n_2 = n_1^2/n$ and obtain $\hat{k}_2$. Then the optimal $k_{opt}$ in (2.39) can be estimated by

$$\hat{k}_{DHPV} = \frac{\hat{k}_1^2}{\hat{k}_2}\left\{\frac{(\log \hat{k}_1)^2}{(2\log n_1 - \log \hat{k}_1)^2}\right\}^{\frac{\log n_1 - \log \hat{k}_1}{\log n_1}}.$$

**Theorem 2.9.** *Under conditions (2.1) and (2.4) with* $A(t)/\left\{t^{1/\alpha}\bar{F}^{-}(t)\right\} = dt^{\rho}$ *for some* $d \neq 0$ *and* $\rho > 0$*, we have*

$$\hat{k}_{DHPV}/k_{opt} \xrightarrow{p} 1 \quad as \quad n \to \infty,$$

*where* $k_{opt}$ *is given in (2.39).*

*Proof.* See Danielsson et al. [26]. □

### 2.2.1.4 Bias Corrected Estimation

A different way for handling this difficult issue of choosing $k$ is to employ a bias corrected estimator. In general, plotting a bias corrected estimator against various $k$'s will give a flat curve over a wider range of $k$, and so choosing $k$ becomes less sensitive. Here we introduce the following three bias corrected estimation procedures.

For simplicity, we assume that as $x \to \infty$,

$$1 - F(x) = cx^{-\alpha} + dx^{-\beta} + o(x^{-\beta}), \tag{2.44}$$

where $c > 0, d \neq 0, \beta > \alpha > 0$. This is a special case of (2.4). Also note that (2.44) implies that

$$\lim_{t\to 0}\frac{(tx)^{1/\alpha}\bar{F}^{-}(tx) - t^{1/\alpha}\bar{F}^{-}(t)}{A(t)} = \frac{x^{\beta/\alpha-1}-1}{\beta/\alpha-1},$$

i.e.,

$$\lim_{t\to 0}\frac{\log \bar{F}^{-}(tx) - \log \bar{F}^{-}(t) + \alpha^{-1}\log x}{A(t)/\left\{t^{1/\alpha}\bar{F}^{-}(t)\right\}} = \frac{x^{\beta/\alpha}-1}{\beta/\alpha-1},$$

where

$$A(t) = \alpha^{-2}(\beta-\alpha)dc^{1-\beta/\alpha}t^{\beta/\alpha-1}. \tag{2.45}$$

**Method 1) Joint estimation.** By noting that $i\log\frac{X_{n,n-i+1}}{X_{n,n-i}}$ can be approximated by an exponential distribution with mean

$$\alpha^{-1}\exp\left\{D_1(i/n)^{\beta_1}\right\}, \quad \text{where} \quad \beta_1 = \beta/\alpha - 1 \quad \text{and} \quad D_1 = -\beta_1 c^{-\beta/\alpha}d,$$

Feuerverger and Hall [38] proposed to estimate $\alpha, \beta, c, d$ simultaneously, which requires that $k$ is a larger order than the optimal one in (2.39). Alternatively, Peng and Qi [82] proposed a censored likelihood estimator which eventually solves the following equations

$$\frac{1}{k}\sum_{i=1}^{k} Q_i^{-1}(\alpha,\beta) = 1 \quad \text{and} \quad \frac{1}{k}\sum_{i=1}^{k} Q_i^{-1}(\alpha,\beta)\log\frac{X_{n,n-i+1}}{X_{n,n-k}} = \beta^{-1} \tag{2.46}$$

for $\beta > \alpha > \hat{\alpha}(k)$, where

$$Q_i(\alpha,\beta) = \frac{\alpha}{\beta}\left(1 + \frac{\alpha\beta}{\alpha-\beta}H(\alpha)\right)\left(\frac{X_{n,n-i+1}}{X_{n,n-k}}\right)^{\beta-\alpha} - \frac{\alpha\beta}{\alpha-\beta}H(\alpha)$$

with

$$H(\alpha) = \frac{1}{\alpha} - \frac{1}{k}\sum_{i=1}^{k}\log\frac{X_{n,n-i+1}}{X_{n,n-k}}.$$

Denote these estimators by $\hat{\alpha}_{PQ}(k)$ and $\hat{\beta}_{PQ}(k)$, and let $\alpha_0$ and $\beta_0$ denote the true values of $\alpha$ and $\beta$, respectively.

**Theorem 2.10.** *Assume that (2.5) holds with $\rho = \beta_0/\alpha_0 - 1$, $\beta_0 > \alpha_0$, $\gamma \geq 0$,*

$$\lim_{t\to 0}\frac{A(t)}{t^{1/\alpha_0}\bar{F}^{-}(t)B(t)} = l_0 \in [-\infty, \infty].$$

*Further assume (2.28),*

$$\sqrt{k}|A(\frac{k}{n})| \to \infty, \quad \sqrt{k}A^2(\frac{k}{n}) \to 0, \quad \sqrt{k}A(\frac{k}{n})B(\frac{k}{n}) \to 0 \quad \text{as} \quad n \to \infty.$$

*Then there exists a solution $(\hat{\alpha}_{PQ}(k), \hat{\beta}_{PQ}(k))^T$ to (2.46) with probability tending to one such that $\hat{\alpha}_{PQ}(k) - \alpha_0 = o_p(1)$, $\hat{\beta}_{PQ}(k) - \beta_0 = o_p(1)$ and*

$$\sqrt{k}\Big(\hat{\alpha}_{PQ}(k) - \alpha_0, \bar{A}(\frac{k}{n})(\hat{\beta}_{PQ}(k) - \beta_0)\Big)^T \xrightarrow{d} N(0, \Sigma),$$

*where $\bar{A}(t) = A(t)/\left\{t^{1/\alpha_0}\bar{F}^{-}(t)\right\}$ and*

$$\Sigma = \begin{pmatrix} \dfrac{\alpha_0^2\beta_0^4}{(\beta_0-\alpha_0)^4} & \dfrac{\alpha_0^2}{\beta_0-\alpha_0} \\ \dfrac{\alpha_0^2}{\beta_0-\alpha_0} & \dfrac{\alpha_0(\beta_0-\alpha_0)^2}{\beta_0^2(2\beta_0-\alpha_0)} \end{pmatrix}.$$

*Proof.* We only prove the asymptotic normality. Hence we assume $\alpha-\alpha_0=o_p(1)$ and $\beta-\beta_0=o_p(1)$ in the following expansions.

Put

$$\bar{B}(t)=B(t)+\bar{A}(t),$$

$$h(x)=\frac{H_{\rho,\gamma}(x)}{1+l_0}-\frac{l_0}{2(1+l_0)}(\frac{x^\rho-1}{\rho})^2,\quad \Delta_1(\delta)=\frac{1}{k}\sum_{i=1}^{k}(\frac{X_{n,n-i+1}}{X_{n,n-k}})^\delta,$$

$$\Delta_2(\delta)=\frac{1}{k}\sum_{i=1}^{k}(\frac{X_{n,n-i+1}}{X_{n,n-k}})^\delta\log\frac{X_{n,n-i+1}}{X_{n,n-k}},$$

$$P_1=\frac{1}{k}\sum_{i=1}^{k}\log\frac{U_{n,i}}{U_{n,k+1}}-\int_0^1\log s\,ds,\quad P_2(\delta)=\frac{1}{k}\sum_{i=1}^{k}(\frac{U_{n,i}}{U_{n,k+1}})^\delta-\int_0^1 s^\delta\,ds$$

and

$$P_3(\delta)=\frac{1}{k}\sum_{i=1}^{k}(\frac{U_{n,i}}{U_{n,k+1}})^\delta\log\frac{U_{n,i}}{U_{n,k+1}}-\int_0^1 s^\delta\log s\,ds.$$

Note that

$$\int_0^1\log s\,ds=-1,\quad \int_0^1 s^\delta\,ds=\frac{1}{\delta+1},\quad \int_0^1 s^\delta\log s\,ds=-\frac{1}{(\delta+1)^2} \tag{2.47}$$

for $\delta>-1$ and

$$\begin{aligned}Q_i^{-1}(\alpha,\beta)=&\frac{\beta}{\alpha}(\frac{X_{n,n-i+1}}{X_{n,n-k}})^{\alpha-\beta}-\frac{\beta^2}{\alpha-\beta}H(\alpha)(\frac{X_{n,n-i+1}}{X_{n,n-k}})^{\alpha-\beta}\\&+\frac{\beta^3}{\alpha(\alpha-\beta)}H(\alpha)(\frac{X_{n,n-i+1}}{X_{n,n-k}})^{2\alpha-2\beta}+O_p\big(H^2(\alpha)\big).\end{aligned} \tag{2.48}$$

Using (2.47) and (2.30), we have

$$\begin{cases}\sqrt{k}P_1=-\int_0^1\frac{W_n(s)}{s}\,ds+W_n(1)+o_p(1),\\ \sqrt{k}P_2(\delta)=-\delta\int_0^1 W_n(s)s^{\delta-1}\,ds+\frac{\delta W_n(1)}{1+\delta}+o_p(1),\\ \sqrt{k}P_3(\delta)=-\int_0^1 W_n(s)\left\{\delta s^{\delta-1}\log s+s^{\delta-1}\right\}ds+\frac{W_n(1)}{(1+\delta)^2},\end{cases}$$

i.e.,

$$\sqrt{k}\big(P_1,P_2(\delta),P_3(\delta)\big)^T\xrightarrow{d}N(0,\Sigma_1), \tag{2.49}$$

where $W_n(s)$ is given in (2.30) and

$$\Sigma_1 = \begin{pmatrix} 1 & -\frac{1}{(1+\delta)^2}+\frac{1}{1+\delta} & \frac{2}{(1+\delta)^3}-\frac{1}{(1+\delta)^2} \\ -\frac{1}{(1+\delta)^2}+\frac{1}{1+\delta} & \frac{1}{(1+2\delta)}-\frac{1}{(1+\delta)^2} & \frac{1}{(1+\delta)^3}-\frac{1}{(1+2\delta)^2} \\ \frac{2}{(1+\delta)^3}-\frac{1}{(1+\delta)^2} & \frac{1}{(1+\delta)^3}-\frac{1}{(1+2\delta)^2} & \frac{1}{(1+2\delta)^3}-\frac{1}{(1+\delta)^4} \end{pmatrix}.$$

By (2.48) and (2.49), we can write (2.46) as

$$\begin{aligned} &\frac{\beta}{\alpha}\Delta_1(\alpha-\beta)-\frac{\beta^2}{\alpha-\beta}H(\alpha)\Delta_1(\alpha-\beta)+\frac{\beta^3}{\alpha(\alpha-\beta)}H(\alpha)\Delta_1(2\alpha-2\beta) \\ &=1+O_p\big(H^2(\alpha)\big) \end{aligned} \tag{2.50}$$

and

$$\begin{aligned} &\frac{\beta}{\alpha}\Delta_2(\alpha-\beta)-\frac{\beta^2}{\alpha-\beta}H(\alpha)\Delta_2(\alpha-\beta)+\frac{\beta^3}{\alpha(\alpha-\beta)}H(\alpha)\Delta_2(2\alpha-2\beta) \\ &=\beta^{-1}+O_p\big(H^2(\alpha)\big). \end{aligned} \tag{2.51}$$

Using (2.11), (2.22), (2.48), (2.49) and Taylor expansions, we can show that, as $n\to\infty$,

$$\begin{aligned} H(\alpha) &= \frac{1}{\alpha}-\frac{1}{\alpha_0}-\Big\{-\alpha_0^{-1}\frac{1}{k}\sum_{i=1}^{k}\log\frac{U_{n,i}}{U_{n,k+1}}+\alpha_0^{-1}\int_0^1\log(s)\,ds \\ &\quad +\bar{A}(\frac{k}{n})\int_0^1\frac{s^\rho-1}{\rho}\,ds+O_p(|\bar{A}(\frac{k}{n})|/\sqrt{k}+|\bar{A}(\frac{k}{n})\bar{B}(\frac{k}{n})|)\Big\} \\ &= \frac{\alpha_0-\alpha}{\alpha_0\alpha}+\alpha_0^{-1}P_1+\frac{\alpha_0}{\beta_0}\bar{A}(\frac{k}{n})+O_p\big(\bar{A}^2(\frac{k}{n})+|\bar{A}(\frac{k}{n})B(\frac{k}{n})|\big), \end{aligned}$$

$$\begin{aligned} \Delta_1(\delta) &= \frac{1}{k}\sum_{i=1}^{k}(\frac{U_{n,i}}{U_{n,k+1}})^{-\delta/\alpha_0}+\delta\bar{A}(\frac{k}{n})\int_0^1 s^{-\delta/\alpha_0}\frac{s^\rho-1}{\rho}\,ds \\ &\quad +O_p\big(\bar{A}^2(\frac{k}{n})+|\bar{A}(\frac{k}{n})B(\frac{k}{n})|\big) \\ &= \frac{\alpha_0}{\alpha_0-\delta}+P_2(-\frac{\delta}{\alpha_0})-\frac{\delta\alpha_0^2}{(\alpha_0-\delta)(\beta_0-\delta)}\bar{A}(\frac{k}{n}) \\ &\quad +O_p\big(\bar{A}^2(\frac{k}{n})+|\bar{A}(\frac{k}{n})B(\frac{k}{n})|\big), \end{aligned}$$

$$\begin{aligned}
\Delta_2(\delta) =& -\frac{\alpha_0}{(\alpha_0-\delta)^2} - \alpha_0^{-1} P_3(-\frac{\delta}{\alpha_0}) + \bar{A}(\frac{k}{n})\Big\{\int_0^1 s^{-\delta/\alpha_0}\frac{s^\rho-1}{\rho}\,ds \\
&-\frac{\delta}{\alpha_0}\int_0^1 \log(s) s^{-\delta/\alpha_0}\frac{x^\rho-1}{\rho}\,ds\Big\} + O_p\big(\bar{A}^2(\frac{k}{n}) + |\bar{A}(\frac{k}{n})B(\frac{k}{n})|\big) \\
=& -\frac{\alpha_0}{(\alpha_0-\delta)^2} - \alpha_0^{-1}P_3(-\frac{\delta}{\alpha_0}) \\
&+ \bar{A}(\frac{k}{n})\left\{-\frac{\alpha_0^2}{(\alpha_0-\delta)(\beta_0-\delta)} - \frac{\delta\alpha_0^2(\alpha_0+\beta_0-2\delta)}{(\beta_0-\delta)^2(\alpha_0-\delta)^2}\right\} \\
&+ O_p\big(\bar{A}^2(\frac{k}{n}) + |\bar{A}(\frac{k}{n})B(\frac{k}{n})|\big) \\
=& -\frac{\alpha_0}{(\alpha_0-\delta)^2} - \alpha_0^{-1}P_3(-\frac{\delta}{\alpha_0}) - \bar{A}(\frac{k}{n})\frac{\alpha_0^2(\alpha_0\beta_0-\delta^2)}{(\alpha_0-\delta)^2(\beta_0-\delta)^2} \\
&+ O_p\big(\bar{A}^2(\frac{k}{n}) + |\bar{A}(\frac{k}{n})B(\frac{k}{n})|\big),
\end{aligned}$$

$$\begin{aligned}
H(\alpha)\Delta_1(\delta) =& \frac{\alpha_0-\alpha}{\alpha(\alpha_0-\delta)} + \frac{1}{\alpha_0-\delta}P_1 + \bar{A}(\frac{k}{n})\frac{\alpha_0^2}{\beta_0(\alpha_0-\delta)} \\
&+ O_p(\frac{|\alpha_0-\alpha|}{\sqrt{k}} + \bar{A}^2(\frac{k}{n}) + |\bar{A}(\frac{k}{n})B(\frac{k}{n})|)
\end{aligned}$$

and

$$\begin{aligned}
H(\alpha)\Delta_2(\delta) =& -\frac{\alpha_0(\alpha_0-\alpha)}{\alpha(\alpha_0-\delta)^2} - \frac{1}{(\alpha_0-\delta)^2}P_1 - \bar{A}(\frac{k}{n})\frac{\alpha_0^2}{\beta_0(\alpha_0-\delta)^2} \\
&+ O_p(\frac{|\alpha_0-\alpha|}{\sqrt{k}} + \bar{A}^2(\frac{k}{n}) + |\bar{A}(\frac{k}{n})B(\frac{k}{n})|)
\end{aligned}$$

for both $\delta = \beta-\alpha$ and $\delta = 2(\beta-\alpha)$. Using these expansions, we can rewrite (2.50) as

$$\begin{aligned}
&(\alpha_0-\alpha)\left\{\frac{\beta-\alpha}{\alpha(\alpha_0-\alpha+\beta)} - \frac{\beta^2}{\alpha(\alpha-\beta)(\alpha_0-\alpha+\beta)}\right. \\
&\left.+\frac{\beta^3}{\alpha^2(\alpha-\beta)(\alpha_0-2\alpha+2\beta)}\right\} \\
&+\frac{\beta}{\alpha}P_2(\frac{\beta-\alpha}{\alpha_0}) - \frac{\beta^2(\alpha_0-2\alpha+\beta)}{\alpha(\alpha_0-\alpha+\beta)(\alpha_0-2\alpha+2\beta)}P_1 \\
&-\left\{\frac{\beta\alpha_0^2(\alpha-\beta)}{\alpha(\alpha_0-\alpha+\beta)(\beta_0-\alpha+\beta)} - \frac{\alpha_0(\alpha_0-\beta_0)}{2\beta_0-\alpha_0}\right\}\bar{A}(\frac{k}{n}) \\
&-\left\{\frac{\beta^2\alpha_0^2}{\beta_0(\alpha-\beta)(\alpha_0-\alpha+\beta)} - \frac{\alpha_0^2}{\alpha_0-\beta_0}\right\}\bar{A}(\frac{k}{n})
\end{aligned}$$

$$+\left\{\frac{\beta^3\alpha_0^2}{\alpha\beta_0(\alpha-\beta)(\alpha_0-2\alpha+2\beta)}-\frac{\alpha_0\beta_0^2}{(\alpha_0-\beta_0)(2\beta_0-\alpha_0)}\right\}\bar{A}(\frac{k}{n})$$
$$=O_p\big(\frac{|\alpha_0-\alpha|}{\sqrt{k}}+\bar{A}^2(\frac{k}{n})+|\bar{A}(\frac{k}{n})B(\frac{k}{n})|\big),$$

which implies that

$$\frac{(\alpha_0-\alpha)(\alpha_0-\beta_0)^3}{\alpha_0^2\beta_0(2\beta_0-\alpha_0)}+\frac{\beta_0}{\alpha_0}P_2(\frac{\beta_0-\alpha_0}{\alpha_0})+\frac{\beta_0(\beta_0-\alpha_0)}{\alpha_0(2\beta_0-\alpha_0)}P_1$$
$$+\bar{A}(\frac{k}{n})(\beta-\beta_0)\left\{\frac{\alpha_0\beta_0}{(2\beta_0-\alpha_0)^2}-\frac{\alpha_0^3}{\beta_0(\alpha_0-\beta_0)^2}\right.$$
$$\left.-\frac{\alpha_0\beta_0(3\alpha_0^2-6\alpha_0\beta_0+2\beta_0^2)}{(\alpha_0-\beta_0)^2(2\beta_0-\alpha_0)^2}\right\}$$
$$=O_p\big(\frac{|\alpha_0-\alpha|}{\sqrt{k}}+\bar{A}^2(\frac{k}{n})+|\bar{A}(\frac{k}{n})B(\frac{k}{n})|+|(\alpha_0-\alpha)\bar{A}(\frac{k}{n})|\big)$$
$$+o_p\big(|(\beta_0-\beta)\bar{A}(\frac{k}{n})|\big),$$

i.e.,

$$(\alpha-\alpha_0)\frac{(\beta_0-\alpha_0)^3}{\alpha_0^2\beta_0^2(2\beta_0-\alpha_0)}+\frac{1}{\alpha_0}P_2(\frac{\beta_0-\alpha_0}{\alpha_0})-\frac{\beta_0-\alpha_0}{\alpha_0(2\beta_0-\alpha_0)}P_1$$
$$-\bar{A}(\frac{k}{n})(\beta-\beta_0)\frac{\alpha_0(\alpha_0-\beta_0)^2}{\beta_0^2(2\beta_0-\alpha_0)^2}$$
$$=O_p\big(\frac{|\alpha_0-\alpha|}{\sqrt{k}}+\bar{A}^2(\frac{k}{n})+|\bar{A}(\frac{k}{n})B(\frac{k}{n})|+|(\alpha_0-\alpha)\bar{A}(\frac{k}{n})|\big)$$
$$+o_p\big(|(\beta_0-\beta)\bar{A}(\frac{k}{n})|\big). \tag{2.52}$$

Similarly we can rewrite (2.51) as

$$(\alpha-\alpha_0)\frac{(\beta_0-\alpha_0)^2(2\alpha_0^2-5\alpha_0\beta_0+\beta_0^2)}{\alpha_0^2\beta_0^3(2\beta_0-\alpha_0)^2}$$
$$-\alpha_0^{-2}P_3(\frac{\beta_0-\alpha_0}{\alpha_0})-\frac{\beta_0^2-3\alpha_0\beta_0+\alpha_0^2}{\alpha_0\beta_0(2\beta_0-\alpha_0)^2}P_1$$
$$-\bar{A}(\frac{k}{n})(\beta-\beta_0)\frac{\alpha_0^2(\alpha_0-\beta_0)(3\beta_0-\alpha_0)}{\beta_0^3(2\beta_0-\alpha_0)^3} \tag{2.53}$$
$$=O_p\big(\frac{|\alpha_0-\alpha|}{\sqrt{k}}+\bar{A}^2(\frac{k}{n})+|\bar{A}(\frac{k}{n})B(\frac{k}{n})|+|(\alpha_0-\alpha)\bar{A}(\frac{k}{n})|\big)$$

$$+o_p\big(|(\beta_0-\beta)\bar{A}(\frac{k}{n})|\big).$$

Solving (2.52) and (2.53) leads to

$$\begin{aligned}\hat{\alpha}_{PQ}(k)-\alpha_0=&\frac{\alpha_0(1+\rho)^4}{\rho^4}P_1-\frac{\alpha_0(1+\rho)^2(2\rho+1)(3\rho+2)}{\rho^5}P_2(\rho)\\&+\frac{\alpha_0(1+\rho)^3(2\rho+1)^2}{\rho^4}P_3(\rho)+o_p(\frac{1}{\sqrt{k}})\end{aligned}\tag{2.54}$$

and

$$\begin{aligned}&\bar{A}(\frac{k}{n})(\hat{\beta}_{PQ}(k)-\beta_0)\\=&\frac{1}{\rho}P_1+\frac{\rho^2-3\rho-2}{\rho^2}P_2(\rho)+\frac{(\rho+1)(2\rho+1)}{\rho}P_3(\rho)+o_p(\frac{1}{\sqrt{k}}).\end{aligned}\tag{2.55}$$

Put

$$D=\begin{pmatrix}\frac{\alpha_0(1+\rho)^4}{\rho^4} & \frac{1}{\rho}\\ -\frac{\alpha_0(1+\rho)^2(2\rho+1)(3\rho+2)}{\rho^5} & \frac{\rho^2-3\rho-2}{\rho^2}\\ \frac{\alpha_0(\rho+1)^3(2\rho+1)^2}{\rho^4} & \frac{(\rho+1)(2\rho+1)}{\rho}\end{pmatrix}.$$

Then the theorem follows from (2.49), (2.54), (2.55) and

$$D^T\Sigma_1 D=\begin{pmatrix}\frac{\alpha_0^2(\rho+1)^4}{\rho^4} & \frac{\alpha_0}{\rho}\\ \frac{\alpha_0}{\rho} & \frac{\rho^2}{(\rho+1)^2(2\rho+1)}\end{pmatrix}=\Sigma. \qquad \square$$

**Remark 2.3.** It turns out that the asymptotic distribution in the above theorem is the same as that in Feuerverger and Hall [38].

**Method 2) Separate estimation.** Gomes and Martins [42] followed the idea of Feuerverger and Hall [38], but proposed to estimate $\alpha$ and other parameters separately instead of jointly, which results in the following bias corrected estimator:

$$\hat{\alpha}_{GM}(k)=\left\{\frac{1}{k}\sum_{i=1}^{k}Z_i-(\frac{1}{k}\sum_{i=1}^{k}i^{\hat{\rho}}Z_i)\frac{(\sum_{i=1}^{k}i^{\hat{\rho}})(\sum_{i=1}^{k}Z_i)-k(\sum_{i=1}^{k}i^{\hat{\rho}}Z_i)}{(\sum_{i=1}^{k}i^{\hat{\rho}})(\sum_{i=1}^{k}i^{\hat{\rho}}Z_i)-k(\sum_{i=1}^{k}i^{2\hat{\rho}}Z_i)}\right\}^{-1},$$

where $Z_i=i\log\frac{X_{n,n-i+1}}{X_{n,n-i}}$ and $\hat{\rho}$ is an external estimator for $\rho$ in (2.4).

**Theorem 2.11.** *Under conditions (2.4) with $\rho>0$ and*

$$k=k(n)\to\infty \quad \textit{and} \quad \sqrt{k}A(k/n)\to\lambda\in\mathbb{R} \quad \textit{as} \quad n\to\infty,$$

*we have*

$$\sqrt{k}\left\{\hat{\alpha}_{GM}(k)-\alpha\right\}\xrightarrow{d}N(0,\frac{\alpha^2(\rho+1)^2}{\rho^2}) \quad \textit{as} \quad n\to\infty.$$

*Proof.* Write

$$\sum_{i=1}^{k}i^{\delta}Z_i=\sum_{i=1}^{k}\left\{i^{\delta+1}-(i-1)^{\delta+1}\right\}\log\frac{X_{n,n-i+1}}{X_{n,n-k}} \quad \text{for any} \quad \delta\ge 0,$$

and put $\bar{A}(t)=\frac{A(t)}{t^{1/\alpha}\bar{F}^{-}(t)}$. Since (2.4) implies (2.20), i.e. (2.38) holds, an application of (2.9) with $f(x)=\log\left\{x^{1/\alpha}\bar{F}^{-}(x)\right\}$ and using (2.30) give

$$\begin{aligned}
&k^{-\delta-1}\sum_{i=1}^{k}i^{\delta}Z_i\\
&=-\alpha^{-1}k^{-\delta-1}\sum_{i=1}^{k}\left\{i^{\delta+1}-(i-1)^{\delta+1}\right\}\log\frac{U_{n,i}}{U_{n,k+1}}\\
&\quad+\bar{A}(\frac{k}{n})k^{-\delta-1}\sum_{i=1}^{k}\left\{i^{\delta+1}-(i-1)^{\delta+1}\right\}\frac{(U_{n,i}/U_{n,k+1})^{\rho}-1}{\rho}\left\{1+o_p(1)\right\}\\
&=-\frac{1+\delta}{\alpha}k^{-1}\sum_{i=1}^{k}(\frac{i}{k})^{\delta}\log\frac{U_{n,i}}{U_{n,k+1}}+O_p(k^{-1})\\
&\quad+\bar{A}(\frac{k}{n})\frac{1+\delta}{k}\sum_{i=1}^{k}(\frac{i}{k})^{\delta}\frac{(U_{n,i}/U_{n,k+1})^{\rho}+1}{\rho}+o_p\big(|\bar{A}(\frac{k}{n})|\big)\\
&=\frac{1}{\alpha(1+\delta)}-\frac{1+\delta}{\alpha}k^{-1}\sum_{i=1}^{k}(\frac{i}{k})^{\delta}\left\{\frac{\frac{n}{i}U_{n,i}}{\frac{n}{k+1}U_{n,k+1}}-1\right\}\\
&\quad+\bar{A}(\frac{k}{n})(1+\delta)\int_0^1 s^{\delta}\frac{s^{\rho}-1}{\rho}\,ds+o_p(|\bar{A}(\frac{k}{n})|+\frac{1}{\sqrt{k}})
\end{aligned}$$

$$
\begin{aligned}
&= \frac{1}{\alpha(1+\delta)} - \frac{1+\delta}{\alpha}\int_0^1 s^{\delta}\left\{\frac{\beta_{n,k}(s)}{s} - \beta_{n,k}(1)\right\} ds - \frac{\bar{A}(k/n)}{1+\rho+\delta} \\
&\quad + o_p\big(|\bar{A}(\frac{k}{n})| + \frac{1}{\sqrt{k}}\big) \\
&= \frac{1}{\alpha(1+\delta)} + \frac{1+\delta}{\alpha}\int_0^1 s^{\delta}\left\{\frac{W_n(s)}{s} - W_n(1)\right\} ds - \frac{\bar{A}(k/n)}{1+\rho+\delta} \\
&\quad + o_p\big(|\bar{A}(\frac{k}{n})| + \frac{1}{\sqrt{k}}\big),
\end{aligned}
$$

which implies that

$$
\begin{aligned}
&\sqrt{k}\left\{\hat{\alpha}_{GM}(k) - \alpha^{-1}\right\} \\
&= \frac{1}{\alpha}\int_0^1 \left\{\frac{W_n(s)}{s} - W_n(1)\right\} ds - \frac{\sqrt{k}\bar{A}(k/n)}{1+\rho} \\
&\quad - \frac{\frac{1}{\alpha(1+\hat{\rho})}\int_0^1\left\{\frac{W_n(s)}{s} - W_n(1)\right\} ds - \frac{\sqrt{k}\bar{A}(k/n)}{(1+\rho)(1+\hat{\rho})} - \frac{1+\hat{\rho}}{\alpha}\int_0^1 s^{\hat{\rho}}\left\{\frac{W_n(s)}{s} - W_n(1)\right\} ds + \frac{\sqrt{k}\bar{A}(k/n)}{1+\rho+\hat{\rho}}}{\alpha(1+\hat{\rho})\left\{\frac{1}{\alpha(1+\hat{\rho})^2} - \frac{1}{\alpha(1+2\hat{\rho})}\right\}} \\
&\quad + o_p(1) \\
&= \frac{1}{\alpha}\left\{1 + \frac{1+2\rho}{\rho^2}\right\}\int_0^1 \left\{\frac{W_n(s)}{s} - W_n(1)\right\} ds \\
&\quad - \frac{(1+\rho)^2(1+2\rho)}{\alpha\rho^2}\int_0^1 s^{\rho}\left\{\frac{W_n(s)}{s} - W_n(1)\right\} ds + o_p(1) \\
&\xrightarrow{d} N(0, \frac{(1+\rho)^2}{\alpha^2\rho^2})
\end{aligned}
$$

by noting that

$$
E\left\{\int_0^1 \big(\frac{W_n(s)}{s} - W_n(1)\big)\, ds\right\}^2 = 1,
$$

$$
E\left\{\int_0^1 s^{\delta}\big(\frac{W_n(s)}{s} - W_n(1)\big)\, ds\right\}^2 = \frac{1}{(1+\delta)^2(1+2\delta)}
$$

and

$$
E\left\{\int_0^1 \big(\frac{W_n(s)}{s} - W_n(1)\big)\, ds\right\}\left\{\int_0^1 s^{\delta}\big(\frac{W_n(s)}{s} - W_n(1)\big)\, ds\right\} = \frac{1}{(1+\delta)^2}.
$$

Hence the theorem holds. □

**Method 3) Direct estimation.** Under (2.4) with $\rho > 0$, by simply comparing the term of $A(k/n)$ in $\hat{\alpha}(k) - \alpha$ to that in $\frac{\hat{\alpha}(k)-2\hat{M}(k)}{\hat{\alpha}(k)}$, where $\hat{M}(k)$ is defined in (2.43), Peng [79] proposed the following bias corrected tail index estimator

$$\hat{\alpha}_P(k) = \hat{\alpha}(k) - \frac{\hat{\alpha}^2(k) - 2\hat{M}(k)}{\hat{\alpha}(k)\hat{\rho}}(1+\hat{\rho}),$$

where $\hat{\rho}$ is a consistent estimator of $\rho$.

**Theorem 2.12.** *Under conditions (2.4) with $\rho > 0$, (2.28) and $\sqrt{k}A(k/n) \to \lambda \in \mathbb{R}$ as $n \to \infty$, we have*

$$\sqrt{k}\left\{\hat{\alpha}_P(k) - \alpha\right\} \xrightarrow{d} N(0, \alpha^2 \frac{1+2\rho+2\rho^2}{\rho^2}) \quad as \quad n \to \infty.$$

*Proof.* The theorem follows from the following expansions, which can be derived by using (2.9) and (2.30) as in the proof of Theorem 2.4:

$$\begin{aligned}\hat{\alpha}(k) &= \alpha - \frac{\alpha}{\sqrt{k}}\int_0^1 \left\{\frac{W_n(s)}{s} - W_n(1)\right\} ds \\ &\quad + \frac{A(k/n)}{(k/n)^{1/\alpha}\bar{F}^{-}(k/n)}\frac{\alpha^2}{1+\rho} + o_p\left(\frac{|A(k/n)|}{(k/n)^{1/\alpha}\bar{F}^{-}(k/n)} + \frac{1}{\sqrt{k}}\right),\end{aligned}$$

$$\begin{aligned}\hat{\alpha}^2(k) &= \alpha^2 - \frac{2\alpha^2}{\sqrt{k}}\int_0^1 \left\{\frac{W_n(s)}{s} - W_n(1)\right\} ds \\ &\quad + \frac{A(k/n)}{(k/n)^{1/\alpha}\bar{F}^{-}(k/n)}\frac{2\alpha^3}{1+\rho} + o_p\left(\frac{|A(k/n)|}{(k/n)^{1/\alpha}\bar{F}^{-}(k/n)} + \frac{1}{\sqrt{k}}\right),\end{aligned}$$

$$\begin{aligned}\hat{M}(k) &= \frac{\alpha^2}{2} + \frac{\alpha^2}{2\sqrt{k}}\int_0^1 \log(s)\left\{\frac{W_n(s)}{s} - W_n(1)\right\} ds \\ &\quad + \frac{A(k/n)}{(k/n)^{1/\alpha}\bar{F}^{-}(k/n)}\frac{\alpha^3(2+\rho)}{2(1+\rho)^2} + o_p\left(\frac{|A(k/n)|}{(k/n)^{1/\alpha}\bar{F}^{-}(k/n)} + \frac{1}{\sqrt{k}}\right),\end{aligned}$$

$$E\left\{\int_0^1 \left(\frac{W_n(s)}{s} - W_n(1)\right) ds\right\}^2 = 1,$$

$$E\left\{\int_0^1 \log(s)\left(\frac{W_n(s)}{s} - W_n(1)\right) ds\right\}^2 = 5,$$

$$E\left\{\int_0^1 \left(\frac{W_n(s)}{s} - W_n(1)\right) ds\right\}\left\{\int_0^1 \log(s)\left(\frac{W_n(s)}{s} - W_n(1)\right) ds\right\} = -2$$

and the fact that $\lim_{t\to 0} t^{1/\alpha}\bar{F}^{-}(t) = c \in (0,\infty)$ implied by $\rho > 0$. □

**Remark 2.4.** The above bias corrected estimator via joint estimation of the first and second order regular variation parameters requires the involved sample fraction $k$ to be a larger order than the optimal choice $k_{opt}$ in (2.39), which results in an estimator with a faster rate of convergence. Another two bias corrected approaches keep the sample fraction at the same order as the optimal choice $k_{opt}$ in (2.39), but have a null asymptotic bias. In general a plot of a bias corrected estimator against $k$'s will show a flat curve for a wider range of $k$, which makes the choice of $k$ less sensitive. Although $\hat{\alpha}_P(k)$ has a larger variance than $\hat{\alpha}_{GM}(k)$, it is less sensitive to the employed external estimator for the second order regular variation parameter $\rho$ in (2.4).

#### 2.2.1.5 Sample Fraction Choice Motivated by Bias Corrected Estimation

Instead of choosing the optimal $k$ in terms of minimizing the asymptotic mean squared error of a tail index estimator, the above bias corrected estimators suggest to first pick up a $k$ as large as possible such that $\sqrt{k}A(k/n) \to \lambda \in \mathbb{R}$, and then to employ a bias corrected estimator with such a chosen $k$. This idea starts with Guillou and Hall [47] and then is generalized by Peng [81] as follows. Under (2.44), let $\hat{\beta}_1$ be a consistent estimator of $\beta_1 = \beta/\alpha - 1$ and choose

$$\begin{aligned}\tilde{k} = \inf\Big\{k : |\sqrt{\ell}(\frac{\hat{\alpha}^2(\ell)}{2\hat{M}(\ell)} - 1)| \ge c_{crit} \text{ for all } \ell \ge k \quad \text{and} \\ \ell \in [n^{\frac{2\hat{\beta}_1}{1+2\hat{\beta}_1}} \wedge (0.01n) + 1, n^{0.99} \vee (n^{\frac{2\hat{\beta}_1}{1+2\hat{\beta}_1}} \log n) \wedge n - 1]\Big\},\end{aligned}$$

where $c_{crit}$ is chosen as 1.25 in practice. Therefore one could employ the above bias corrected estimators $\hat{\alpha}_{GM}(\tilde{k})$ or $\hat{\alpha}_P(\tilde{k})$ to estimate $\alpha$. However the asymptotic limit of both $\sqrt{\tilde{k}}\left\{\hat{\alpha}_{GM}(\tilde{k}) - \alpha\right\}$ and $\sqrt{\tilde{k}}\left\{\hat{\alpha}_P(\tilde{k}) - \alpha\right\}$ is no longer a normal distribution since $\tilde{k}/n^{\frac{2\beta_1}{1+2\beta_1}}$ converges in distribution to a stopping time rather than converges in probability to a constant.

### 2.2.2 Other Tail Index Estimators

**i)** *Kernel estimators in Csörgő et al. [24].* Let $K(u)$ denote a nonnegative, nonincreasing and right continuous function satisfying

$$\int_0^\infty K(u)\,du = 1 \text{ and } \int_0^\infty u^{-1/2}K(u)\,du < \infty.$$

Let $\lambda = \lambda_n > 0$ be a bandwidth satisfying $\lambda \to 0$ and $n\lambda \to \infty$ as $n \to \infty$. Then a class of kernel estimators for $\alpha$ in (2.1) is defined as

$$\begin{aligned}\hat{\alpha}_{CDM}(\lambda) &= \frac{\int_0^{1/\lambda} K(u)\,du}{\int_0^{1/\lambda} \log^+ \bar{F}^-(Q_n(u\lambda))\,d\{uK(u)\}}\\ &= \frac{\int_0^{1/\lambda} K(u)\,du}{\sum_{j=1}^n \frac{j}{n\lambda}K(\frac{j}{n\lambda})(\log^+ X_{n,n-j+1} - \log^+ X_{n,n-j})},\end{aligned}$$

where $\log^+(x) = \log(\max(x, 1))$ and $Q_n(s)$ is defined in (2.27).

**Theorem 2.13.** *In addition to the above conditions on $K$ and $\lambda$, we further assume (2.1), (2.4) with $\rho > 0$,*

$$\sqrt{n\lambda}\lambda^{-1}K(1/\lambda) \to 0, \quad \sqrt{n\lambda}A(\lambda) \to 0 \quad as \quad n \to \infty.$$

*Then*

$$\sqrt{n\lambda}\left\{\hat{\alpha}_{CDM}(\lambda) - \alpha\right\} \xrightarrow{d} N(0, \frac{\alpha^2}{\int_0^\infty K^2(u)\,du}) \quad as \quad n \to \infty.$$

*Proof.* Since $\rho > 0$, we have $\log \bar{F}^-(t) = \log c - \log(t)/\alpha + O(|A(t)|)$ for some $c > 0$ as $t \to 0$. Like the proof of Theorem 2.4, it follows from (2.9), (2.27) and Lemma 2 in Csörgő et al. [24] that

$$\begin{aligned}&\sqrt{n\lambda}\left\{\int_0^{1/\lambda} \log^+ \bar{F}^-(Q_n(u\lambda))\,d\{uK(u)\} - \alpha^{-1}\int_0^{1/\lambda} K(u)\,du\right\}\\ &= \sqrt{n\lambda}\int_0^{1/\lambda} \left\{\log((Q_n(u\lambda))^{1/\alpha}\bar{F}^-(Q_n(u\lambda))) - \log(\lambda^{1/\alpha}\bar{F}^-(\lambda))\right\}\,d\{uK(u)\}\\ &\quad + \sqrt{n\lambda}\int_0^{1/\lambda} \left\{-\alpha^{-1}\log Q_n(u\lambda) + \alpha^{-1}\log(u\lambda)\right\}\,d\{uK(u)\}\\ &\quad + \sqrt{n\lambda}\left\{\int_0^{1/\lambda} \big(-\alpha^{-1}\log u + \log \bar{F}^-(\lambda)\big)\,d\{uK(u)\} - \alpha^{-1}\int_0^{1/\lambda} K(u)\,du\right\}\\ &= o_p(1) - \alpha^{-1}\sqrt{n\lambda}\int_0^{1/\lambda} \frac{Q_n(u\lambda) - u\lambda}{u\lambda}\,d\{uK(u)\}\end{aligned}$$

$$
\begin{aligned}
&+\sqrt{n\lambda}\left\{\int_0^{1/\lambda}\left(-\alpha^{-1}\log u+\log c-\alpha^{-1}\log\lambda\right)d\left\{uk(u)\right\}-\alpha^{-1}\int_0^{1/\lambda}K(u)\,du\right\}\\
&=\alpha^{-1}\int_0^{1/\lambda}\frac{B_n(u\lambda)}{u\sqrt{\lambda}}\,d\{uK(u)\}+\sqrt{n\lambda}\log(c)\lambda^{-1}K(1/\lambda)+o_p(1)\\
&=\alpha^{-1}\left\{\int_0^{1/\lambda}\int_0^{1/\lambda}\frac{\min(u,v)}{uv}\,d\big(uK(u)\big)d\big(vK(v)\big)\right\}^{1/2}N(0,1)+o_p(1),
\end{aligned}
$$

which converges in distribution to $N(0,\alpha^{-2}\int_0^\infty K^2(u)\,du)$, i.e., the theorem holds. □

**ii)** *Linear combinations of intermediate order statistics in Viharos [102].* Let

$$
\left\{d_{n,n-i+1}\right\}_{i=1}^n,\quad \left\{e_{n,n-i+1}\right\}_{i=1}^n,\quad -\infty<\delta<\infty,\ 0<a<b<\infty,
$$

and $m>0$ be known, and define

$$
K_\delta(a,b,m)=\begin{cases}\dfrac{1}{(1+\delta)^2}\left\{\log\dfrac{a^a}{b^b}-\log\dfrac{(a+m)^{a+m}}{(b+m)^{b+m}}\right\} & \text{if}\quad \delta\neq-1,\\ m(b-a) & \text{if}\quad \delta=-1.\end{cases}
$$

Then a class of estimators for $\alpha$ in (2.1) is defined as

$$
\begin{aligned}
\hat\alpha_V(k)=&\frac{k^{\delta+1}K_\delta(a,b,m)}{n^\delta}\left\{\sum_{i=[a^{1/(1+\delta)}k]+1}^{[(a+m)^{1/(1+\delta)}k]}d_{n,n-i+1}\log^+X_{n,n-i+1}\right.\\
&\left.-\sum_{i=[b^{1/(1+\delta)}k]+1}^{[(b+m)^{1/(1+\delta)}k]}e_{n,n-i+1}\log^+X_{n,n-i+1}\right\}^{-1}.
\end{aligned}
$$

To derive the asymptotic distribution, we need the following conditions:

- $d_{n,i}=n\int_{(i-1)/n}^{i/n}\bar L_1(t)\,dt$ and $e_{n,i}=\int_{(i-1)/n}^{i/n}\bar L_2(t)\,dt$ for $1\le i\le n$ and some nonnegative continuous functions $\bar L_1$ and $\bar L_2$ defined on $(0,1)$.
- For $i=1,2$, there exists a constant $0<\mu_i<1$ such that the function $\bar L_i$ is Lipschitz on $[1-\mu_i,1-\mu_0]$ for all $0<\mu_0<\mu_i$, and $\bar L_i(1-t)=t^\delta\bar l_i(t)$ on $(0,1)$, where $\bar l_i(t)\in RV_0^0$, $\lim_{t\to0}\bar l_i(t)=1$ and $\bar l_i'(t)=t^{-1}\bar l_i(t)\epsilon_i(t)$ on $(0,\mu_0)$ with a continuous function $\epsilon_i(t)$ for which $\lim_{t\to0}\epsilon_i(t)=0$.

- (2.1) and (2.28) hold with $\log(n)/\sqrt{k} \to 0$ and

$$\lim_{n\to\infty} \sqrt{k}(\frac{n}{k})^{\delta+1} \int_{[x^{1/(\delta+1)}k]/n}^{[(x+m)^{1/(\delta+1)}k]/n} s^{\delta}|y(s)-1|\log(s)\,ds = 0$$

for $(x, y)^T = (a, \bar{l}_1)^T, (b, \bar{l}_2)^T$.

**Theorem 2.14.** *Under the above conditions, we have*

$$k^{1/2}K_{\delta}(a,b,m)\left\{\hat{\alpha}_V^{-1}(k) - \alpha^{-1} + Q_n\right\}$$

$$\xrightarrow{d} \alpha^{-1}\left\{\int_{a^{1/(\delta+1)}}^{(a+m)^{1/(\delta+1)}} W(x)x^{\delta-1}\,dx - \int_{b^{1/(\delta+1)}}^{(b+m)^{1/(\delta+1)}} W(x)x^{\delta-1}\,dx\right\}$$

*as $n \to \infty$, where $W(s)$ is a standard Wiener process and*

$$Q_n = \frac{-n^{\delta+1}/k^{\delta+1}}{K_{\delta}(a,b,m)}\left\{\int_{a^{1/(\delta+1)}k/n}^{(a+m)^{1/(\delta+1)}k/n} s^{\delta}\log\left(\bar{F}(s)s^{\alpha}\right)ds \right.$$
$$\left. - \int_{b^{1/(\delta+1)}k/n}^{(b+m)^{1/(\delta+1)}k/n} s^{\delta}\log\left(\bar{F}(s)s^{\alpha}\right)ds\right\}.$$

*Proof.* See Theorem 1 of Viharos [102]. □

**iii)** *Least squares estimators in Csörgő and Viharos [23].* Define

$$\hat{\alpha}_{CV}^{(1)}(k) = \left\{\frac{1}{k}\sum_{i=1}^{k}\log(\frac{n}{i})\log(X_{n,n-i+1}) - \frac{1}{k^2}\sum_{i=1}^{k}\log(X_{n,n-i+1})\sum_{i=1}^{k}\log(\frac{n}{i})\right\}^{-1}$$
$$\times\left\{\frac{1}{k}\sum_{i=1}^{k}\log^2(\frac{n}{i}) - (\frac{1}{k}\sum_{i=1}^{k}\log(\frac{n}{i}))^2\right\},$$

$$\hat{\alpha}_{CV}^{(2)}(k) = \left\{\sum_{i=1}^{k}\log(\frac{n}{i})\log(X_{n,n-i+1})\right\}^{-1}\left\{\sum_{i=1}^{k}\log(\frac{n}{i})\right\},$$

$$\hat{\alpha}_{CV}^{(3)}(k) = \left\{\frac{1}{k}\sum_{i=1}^{k}\log^2(X_{n,n-i+1}) - (\frac{1}{k}\sum_{i=1}^{k}\log(X_{n,n-i+1}))^2\right\}^{-1}$$
$$\times\left\{\frac{1}{k}\sum_{i=1}^{k}\log(\frac{n}{i})\log(X_{n,n-i+1}) - \frac{1}{k^2}\sum_{i=1}^{k}\log(X_{n,n-i+1})\sum_{i=1}^{k}\log(\frac{n}{i})\right\}.$$

**Theorem 2.15.** *Suppose (2.1) and (2.28) hold.*
*A) If $k/\log^4(n) \to \infty$, then as $n \to \infty$*

$$\sqrt{k}\left\{\hat{\alpha}_{CV}^{(1)}(k) - \mu_n^{(1)}\right\} \xrightarrow{d} N(0, 2\alpha^2)$$

*and*

$$\sqrt{k}\left\{\hat{\alpha}_{CV}^{(3)}(k) - \mu_n^{(3)}\right\} \xrightarrow{d} N(0, 2\alpha^2),$$

*where*

$$\mu_n^{(1)} = -\frac{1}{\frac{n}{k}\int_0^{k/n} \log\left(\bar{F}^{-}(t)\right)\left\{1 + \log(\frac{nt}{k})\right\} dt}$$

*and*

$$\mu_n^{(3)} = \frac{\mu_n^{(1)}}{(n/k)\int_0^{k/n} \log^2\left(\bar{F}(t)\right) dt - (n/k)^2\left\{\int_0^{k/n} \log\left(\bar{F}(t)\right) dt\right\}^2}.$$

*B) If $\left\{k\log^2(n/k)\right\}/\log^4 n \to \infty$, then*

$$\sqrt{k}\log(\frac{n}{k})\left\{\hat{\alpha}_{CV}^{(2)}(k) - \mu_n^{(2)}\right\} \xrightarrow{d} N(0, 2\alpha^2) \quad as \quad n \to \infty,$$

*where*

$$\mu_n^{(2)} = -\frac{\int_0^{k/n} \log^2 t\, dt}{\int_0^{k/n} \log\left(\bar{F}^{-}(t)\right)\log(t)\, dt}.$$

*Proof.* See Csörgő and Viharos [23]. □

**Remark 2.5.** In the above theorem, we have $\mu_n^{(i)} \to \alpha$ for $i = 1, 2, 3$ as $n \to \infty$. In order to replace $\mu_n^{(i)}$ by $\alpha$ in the above theorem, a second order regular variation condition as (2.4) could be employed, which will guide the choice of $k$ as well.

## 2.3 HIGH QUANTILE ESTIMATION

Quantile at level $1 - p \in (0, 1)$ is defined as $F^{-}(1 - p) = \bar{F}^{-}(p)$, which plays an important role in robust statistics and model diagnostics such as QQ-plot, and has important applications in risk management such as Value-at-Risk. Nonparametric estimation of a quantile and its asymptotic distribution depend on whether the level $1 - p$ is fixed or intermediate or extreme.

Consider the case of upper quantile and let $n$ denote the sample size. Then an intermediate quantile means $p = p_n \to 0$ and $np_n \to \infty$ as $n \to \infty$, and an extreme/high quantile means $p = p_n \to 0$ and $np_n \to c \in [0, \infty)$ as $n \to \infty$. Obviously, when $c = 0$, empirical quantile seriously underestimates this high quantile. This section demonstrates how extreme value theory can be employed to improve the high quantile estimation of a heavy tailed distribution.

Assume $X_1, \cdots, X_n$ are independent and identically distributed random variables with distribution function $F$ satisfying (2.1). Since $\bar{F}^{-}(p)/\bar{F}^{-}(k/n) \sim (np/k)^{-1/\alpha}$ and $\bar{F}^{-}(k/n)$ can be estimated by the empirical quantile $X_{n,n-k}$ nonparametrically when $k$ satisfies (2.28), one can estimate the high quantile $x_p := \bar{F}^{-}(p)$ with $p = p_n \to 0$ by

$$\hat{x}_p = X_{n,n-k}(np/k)^{-1/\hat{\alpha}(k)}, \tag{2.56}$$

where $\hat{\alpha}(k)$ is the Hill estimator defined in (2.31).

**Theorem 2.16.** *Under conditions (2.4), (2.28),*

$$\frac{\sqrt{k}A(k/n)}{(k/n)^{1/\alpha}\bar{F}^{-}(k/n)} \to \lambda \in \mathbb{R}, \quad p = p_n \to 0, \quad \frac{np}{k} \to 0, \quad \log(np) = o(\sqrt{k}),$$

*we have*

$$\frac{\sqrt{k}}{\log(np/k)}\left\{\frac{\hat{x}_p}{x_p} - 1\right\} \xrightarrow{d} N\left(\frac{\lambda}{1+\rho}, \frac{1}{\alpha^2}\right) \quad as \quad n \to \infty.$$

*Proof.* Like the proof of Theorem 2.4, we can show that

$$\begin{aligned}\frac{\hat{x}_p}{x_p} - 1 &= \left\{\frac{\bar{F}^{-}(U_{n,k+1})}{\bar{F}^{-}(p)} - \left(\frac{p}{U_{n,k+1}}\right)^{1/\alpha}\right\}\left(\frac{np}{k}\right)^{-1/\hat{\alpha}(k)} \\ &\quad + \left\{\left(\frac{p}{U_{n,k+1}}\right)^{1/\alpha} - \left(\frac{np}{k+1}\right)^{1/\alpha}\right\}\left(\frac{np}{k}\right)^{-1/\hat{\alpha}(k)} + \left(\frac{np}{k}\right)^{1/\alpha - 1/\hat{\alpha}(k)} - 1 \\ &= \log\left(\frac{np}{k}\right)\left\{\frac{1}{\alpha} - \frac{1}{\hat{\alpha}(k)}\right\}\{1 + o_p(1)\},\end{aligned}$$

which implies the theorem by using the asymptotic distribution of $\hat{\alpha}(k)$ in Theorem 2.4. □

By minimizing the asymptotic mean squared error of $\hat{x}_p$ in the above theorem, one obtains the optimal choice of $k$, which happens to be the same as the optimal choice in $\hat{\alpha}(k)$ given in (2.39) when $A(t)/\left\{t^{1/\alpha}\bar{F}^{-}(t)\right\} = dt^{\rho}$ for some $d \neq 0$ and $\rho > 0$. Therefore, both data-driven methods for choosing the sample fraction in the Hill estimator and some bias-corrected techniques for the Hill estimator can be employed directly for the above high quantile estimation. Some details are available in Gomes and Pestana [43].

## 2.4 EXTREME TAIL PROBABILITY ESTIMATION

A closely related problem to the high quantile estimation is to estimate an extreme tail probability for a heavy tailed loss variable, i.e., estimate $\bar{F}(x_0)$ for a large $x_0$. By noting that

$$\bar{F}(x_0) = \frac{\bar{F}(\frac{x_0}{X_{n,n-k}} X_{n,n-k})}{\bar{F}(X_{n,n-k})} \bar{F}(X_{n,n-k}) \sim \frac{k}{n}(\frac{x_0}{X_{n,n-k}})^{-\alpha} \text{ when } k \text{ satisfies (2.28)},$$

one can estimate $\bar{F}(x_0)$ by

$$\hat{\bar{F}}(x_0) = \frac{k}{n}(\frac{x_0}{X_{n,n-k}})^{-\hat{\alpha}(k)},$$

where $\hat{\alpha}(k)$ is the Hill estimator given in (2.31). This extreme tail probability estimator was studied by Hall and Weissman [53].

**Theorem 2.17.** *Under conditions (2.4), (2.28), $x_0/\bar{F}^{-}(k/n) \to \infty$,*

$$\sqrt{k}\frac{A(k/n)}{(k/n)^{1/\alpha}\bar{F}^{-}(k/n)} \to \lambda \in \mathbb{R}, \quad \frac{\sqrt{k}}{\log(x_0/\bar{F}^{-}(k/n))} \to \infty,$$

*we have*

$$\frac{\sqrt{k}}{\log(x_0/\bar{F}^{-}(k/n))}\left\{\frac{\hat{\bar{F}}(x_0)}{\bar{F}(x_0)} - 1\right\} \xrightarrow{d} N(-\frac{\lambda\alpha^2}{1+\rho}, \alpha^2) \quad \text{as} \quad n \to \infty.$$

*Proof.* Write

$$\begin{aligned}\bar{F}(x_0) - \frac{k}{n}(\frac{x_0}{X_{n,n-k}})^{-\alpha} &= \left\{\frac{\bar{F}(\frac{x_0}{X_{n,n-k}} X_{n,n-k})}{\bar{F}(X_{n,n-k})} - (\frac{x_0}{X_{n,n-k}})^{-\alpha}\right\} \bar{F}(X_{n,n-k}) \\ &\quad + \frac{k}{n}(\frac{x_0}{X_{n,n-k}})^{-\alpha}\left\{\frac{n}{k}\bar{F}(X_{n,n-k}) - 1\right\}\end{aligned}$$

and

$$\hat{\bar{F}}(x_0) - \frac{k}{n}(\frac{x_0}{X_{n,n-k}})^{-\alpha} = -\frac{k}{n}(\frac{x_0}{X_{n,n-k}})^{-\alpha}\log(\frac{x_0}{X_{n,n-k}})\left\{\hat{\alpha}(k)-\alpha\right\}\left\{1+o_p(1)\right\}.$$

Then we have

$$\frac{\hat{\bar{F}}(x_0)}{\bar{F}(x_0)} - 1 = -\log(\frac{x_0}{X_{n,n-k}})\left\{\hat{\alpha}(k)-\alpha\right\}\left\{1+o_p(1)\right\},$$

which implies the theorem by using the asymptotic distribution of $\hat{\alpha}(k)$ in Theorem 2.4. □

**Remark 2.6.** It follows from the above theorem that the optimal choice of $k$ in terms of minimizing the asymptotic mean squared error of $\hat{\bar{F}}(x_0)$ is the same as that for the Hill estimator, i.e., $k_{opt}$ given in (2.39). Hence both data-driven methods for choosing $k_{opt}$ in (2.39) and bias corrected tail index estimators replacing $\hat{\alpha}(k)$ in $\hat{\bar{F}}(x_0)$ can be employed for estimating an extreme tail probability.

## 2.5 INTERVAL ESTIMATION

### 2.5.1 Confidence Intervals for Tail Index

For constructing a confidence interval for the tail index $\alpha$ in (2.1) from a random sample $X_1, \cdots, X_n$, we introduce the following three methods by focusing on the Hill estimator $\hat{\alpha}(k)$ given in (2.31).

#### *2.5.1.1 Normal Approximation Method*

Like nonparametric smoothing estimation, one should undersmooth $\hat{\alpha}(k)$ for the purpose of interval estimation; see Hall [50]. That is, one chooses $k$ such that $\lambda = 0$ in Theorem 2.4 and then construct a confidence interval for $\alpha$ based on

$$\sqrt{k}\left\{\hat{\alpha}(k)/\alpha - 1\right\} \xrightarrow{d} N(0,1) \quad \text{as} \quad n\to\infty.$$

In order to evaluate the accuracy of such an interval and choose an optimal $k$ in terms of coverage accuracy, Cheng and Peng [20] derived the following theorem by extending the result in Cheng and Pan [19]. Other expansions are available in Haeusler and Segers [48].

**Theorem 2.18.** *Under (2.44) and (2.28),*

$$P\Big(\sqrt{k}\Big\{\frac{\alpha}{\hat{\alpha}(k)}-1\Big\}\le x\Big)=\Phi(x)+\phi(x)\Big\{\frac{1-x^2}{3\sqrt{k}}+\frac{d(\beta-\alpha)}{\beta c^{\beta/\alpha}}\sqrt{k}(\frac{n}{k})^{-\frac{\beta}{\alpha}+1}\Big\}$$
$$+o\Big(\frac{1}{\sqrt{k}}+\sqrt{k}(\frac{n}{k})^{-\beta/\alpha+1}\Big)$$

*holds uniformly for $x\in\mathbb{R}$, where $\Phi(x)$ and $\phi(x)$ denote the distribution function and density function of a standard normal random variable respectively.*

*Proof.* See Cheng and Peng [20]. □

### 2.5.1.2 Bootstrap Method

It is known that a full-sample bootstrap method fails for extremes (see Angus [3]) and can not catch the asymptotic bias of a tail index estimator (see Hall [51]). Since constructing a confidence interval for a tail index usually undersmoothes first, El-Nouty and Guillou [35] showed that a full-sample bootstrap method is applicable when $\lambda=0$ in Theorem 2.4.

### 2.5.1.3 Empirical Likelihood Method

Since Owen [76,77] introduced the so-called empirical likelihood method for constructing a confidence interval for a mean and a confidence region for a mean vector, researchers have proved that this method is very efficient in interval estimation and hypothesis tests with applications to various fields; see Owen [78] for an overview on the empirical likelihood method. A useful way in formulating an empirical likelihood function is via estimating equations as developed by Qin and Lawless [89].

In order to introduce proper score equations for an empirical likelihood of the tail index $\alpha$ in (2.1), we employ the censored likelihood function in Section 2.2.1. That is, by assuming that all data are subject to left-censoring at $T$, the censored likelihood function has been given in (2.32). With $T$ replaced by $X_{n,n-k}$ we have the following censored likelihood function

$$\prod_{i=1}^{n}\{c\alpha X_i^{-\alpha-1}\}^{\delta_i}\Big\{1-cX_{n,n-k}^{-\alpha}\Big\}^{1-\delta_i},\quad\text{where } \delta_i=I(X_i>X_{n,n-k}),\qquad(2.57)$$

which results in the score equations

$$\begin{cases} \sum_{i=1}^{n}\left\{\frac{\delta_i}{c}-\frac{(1-\delta_i)X_{n,n-k}^{-\alpha}}{1-cX_{n,n-k}^{-\alpha}}\right\}=0, \\ \sum_{i=1}^{n}\left\{\delta_i(\frac{1}{\alpha}-\log X_i)+\frac{(1-\delta_i)cX_{n,n-k}^{-\alpha}\log X_{n,n-k}}{1-cX_{n,n-k}^{-\alpha}}\right\}=0. \end{cases}$$

Hence, using the empirical likelihood method based on estimating equations in Qin and Lawless [89], we define the empirical likelihood function as $\max\left\{\prod_{i=1}^{n}(np_i)\right\}$ subject to

$$p_1\geq 0,\cdots,p_n\geq 0,\sum_{i=1}^{n}p_i=1,\quad \sum_{i=1}^{n}p_i\left\{\frac{\delta_i}{c}-\frac{(1-\delta_i)X_{n,n-k}^{-\alpha}}{1-cX_{n,n-k}^{-\alpha}}\right\}=0,$$

$$\text{and}\quad \sum_{i=1}^{n}p_i\left\{\delta_i(\frac{1}{\alpha}-\log X_i)+\frac{(1-\delta_i)cX_{n,n-k}^{-\alpha}\log X_{n,n-k}}{1-cX_{n,n-k}^{-\alpha}}\right\}=0.$$

After a simplification, the empirical likelihood function for $\alpha$ becomes

$$L(\alpha)=\sup\left\{\prod_{i=1}^{n}(np_i):p_1\geq 0,\cdots,p_n\geq 0,\sum_{i=1}^{n}p_i=1,\sum_{i=1}^{n}p_i\delta_iZ_i(\alpha)=0\right\},$$

where $Z_i(\alpha)=\frac{1}{\alpha}-\log\frac{X_i}{X_{n,n-k}}$. By the Lagrange multiplier technique we have

$$p_i=\frac{1}{n\big(1+\lambda\delta_iZ_i(\alpha)\big)}$$

and

$$l(\alpha):=-2\log L(\alpha)=2\sum_{i=1}^{n}\log\{1+\lambda\delta_iZ_i(\alpha)\}=2\sum_{i=1}^{k}\log\left\{1+\lambda\tilde{Z}_i(\alpha)\right\},$$

where $\tilde{Z}_i(\alpha)=\alpha^{-1}-\log(X_{n,n-i+1}/X_{n,n-k})$ and $\lambda=\lambda(\alpha)$ satisfies

$$0=\sum_{i=1}^{n}\frac{\delta_iZ_i(\alpha)}{1+\lambda\delta_iZ_i(\alpha)}=\sum_{i=1}^{k}\frac{\tilde{Z}_i(\alpha)}{1+\lambda\tilde{Z}_i(\alpha)}. \tag{2.58}$$

**Remark 2.7.** The log-empirical likelihood statistic $l(\alpha)=-2\log L(\alpha)$ depends only on $\tilde{Z}_i(\alpha)$, $1\leq i\leq k$. As a matter of fact, we can directly work with $\tilde{Z}_i(\alpha)$, $1 \le i \le k$ by defining the empirical likelihood function for $\alpha$ as

$$L(\alpha) = \sup\left\{\prod_{i=1}^{k}(kp_i) : p_1 \ge 0, \cdots, p_k \ge 0, \sum_{i=1}^{k} p_i = 1, \sum_{i=1}^{k} p_i \tilde{Z}_i(\alpha) = 0\right\}.$$

Then it follows from the standard procedure for the empirical likelihood method that $l(\alpha) = 2\sum_{i=1}^{k} \log\left\{1 + \lambda \tilde{Z}_i(\alpha)\right\}$, where $\lambda$ is the solution to Eq. (2.58).

**Theorem 2.19.** *Under conditions (2.4), (2.28) and*

$$\lim_{n\to\infty} \sqrt{k}\frac{A(k/n)}{(k/n)^{1/\alpha}\bar{F}^{-}(k/n)} = 0,$$

*$l(\alpha_0)$ converges in distribution to a chi-squared limit with one degree of freedom as $n \to \infty$, where $\alpha_0$ is the true value of $\alpha$.*

*Proof.* It follows from (2.9), (2.30) and similar arguments in proving Theorem 2.4 that

$$\frac{1}{\sqrt{k}}\sum_{i=1}^{k} \tilde{Z}_i(\alpha_0) \xrightarrow{d} N(0, \frac{1}{\alpha_0^2}), \quad \frac{1}{k}\sum_{i=1}^{k} \tilde{Z}_i^2(\alpha_0) \xrightarrow{p} \frac{1}{\alpha_0^2}, \quad \max_{1\le i\le k} |\tilde{Z}_i(\alpha_0)| = o_p(\sqrt{k}). \tag{2.59}$$

Since

$$0 = \sum_{i=1}^{k} \frac{\tilde{Z}_i(\alpha_0)}{1 + \lambda\tilde{Z}_i(\alpha_0)} = \sum_{i=1}^{k} \tilde{Z}_i(\alpha_0)\left\{1 - \frac{\lambda\tilde{Z}_i(\alpha_0)}{1 + \lambda\tilde{Z}_i(\alpha_0)}\right\},$$

we have

$$\begin{aligned}|\sum_{i=1}^{k} \tilde{Z}_i(\alpha_0)| &= |\sum_{i=1}^{k} \frac{\lambda\tilde{Z}_i^2(\alpha_0)}{1 + \lambda\tilde{Z}_i(\alpha_0)}| = |\lambda| \sum_{i=1}^{k} \frac{\tilde{Z}_i^2(\alpha_0)}{1 + \lambda\tilde{Z}_i(\alpha_0)} \\ &\ge |\lambda| \sum_{i=1}^{k} \frac{\tilde{Z}_i^2(\alpha_0)}{1 + |\lambda| \max_{1\le i\le k} |\tilde{Z}_i(\alpha_0)|} \\ &= \frac{|\lambda|}{1 + |\lambda| \max_{1\le i\le k} |\tilde{Z}_i(\alpha_0)|} \sum_{i=1}^{k} \tilde{Z}_i^2(\alpha_0),\end{aligned}$$

i.e.,

$$|\sum_{i=1}^{k} \tilde{Z}_i(\alpha_0)| \ge |\lambda| \left\{\sum_{i=1}^{k} \tilde{Z}_i^2(\alpha_0) - \left(\max_{1\le i\le k} |\tilde{Z}_i(\alpha_0)|\right) | \sum_{i=1}^{k} \tilde{Z}_i(\alpha_0)|\right\},$$

i.e.,

$$
\begin{aligned}
&|\frac{1}{\sqrt{k}}\sum_{i=1}^{k}\tilde{Z}_i(\alpha_0)| \\
&\geq \sqrt{k}|\lambda|\left\{\frac{1}{k}\sum_{i=1}^{k}\tilde{Z}_i^2(\alpha_0)-(\frac{1}{\sqrt{k}}\max_{1\leq i\leq k}|\tilde{Z}_i(\alpha_0)|)|\frac{1}{\sqrt{k}}\sum_{i=1}^{k}\tilde{Z}_i(\alpha_0)|\right\},
\end{aligned}
$$

which implies that

$$|\lambda| = O_p(\frac{1}{\sqrt{k}}) \tag{2.60}$$

by using (2.59). Further we can show that

$$\lambda = \frac{\sum_{i=1}^{k}\tilde{Z}_i(\alpha_0)}{\sum_{i=1}^{k}\tilde{Z}_i^2(\alpha_0)} + o_p(\frac{1}{\sqrt{k}}). \tag{2.61}$$

Therefore, by noting that $\max_{1\leq i\leq k}|\lambda\tilde{Z}_i(\alpha_0)| = o_p(1)$, which is implied by (2.59) and (2.60), it follows from (2.59)–(2.61) and Taylor expansions that

$$
\begin{aligned}
l(\alpha_0) &= 2\sum_{i=1}^{k}\lambda\tilde{Z}_i(\alpha_0) - \sum_{i=1}^{k}\lambda^2\tilde{Z}_i^2(\alpha_0) + o_p(1) = \frac{(\sum_{i=1}^{k}\tilde{Z}_i(\alpha_0))^2}{\sum_{i=1}^{k}\tilde{Z}_i^2(\alpha_0)} + o_p(1) \\
&\xrightarrow{d} \chi^2(1) \quad \text{as} \quad n \to \infty.
\end{aligned}
$$

This completes the proof. □

Based on the above theorem (often called **Wilks Theorem**), an empirical likelihood confidence interval for $\alpha_0$ with level $\xi$ is

$$I_\xi = \left\{\alpha : l(\alpha) \leq \chi^2_{1,\xi}\right\}, \tag{2.62}$$

where $\chi^2_{1,\xi}$ denotes the $\xi$-th quantile of a chi-squared distribution with one degree of freedom. Furthermore, it follows from the arguments in Hall and La Scala [52] that $l(\alpha)$ can be shown to be a convex function of $\alpha > 0$, i.e., $I_\xi$ is an interval.

**Remark 2.8.** In order to improve the accuracy of the coverage probability for confidence interval $I_\xi$ defined in (2.62) when $k$ is small, Peng and Qi [84] recommended to use different quantiles other than $\chi^2_{1,\xi}$. Those quantiles depend on $k$ and are derived from distributions of the empirical likelihood statistic based on the standard exponential distribution. See Table A.1 in Appendix A for some estimated quantiles from simulation when $1-\xi = 0.10$, 0.05 and 0.01 for $10 \leq k \leq 99$.

## 2.5.2 Confidence Intervals for High Quantile

Like the study of intervals for a tail index, one could employ a normal approximation method, a full-sample bootstrap method and an empirical likelihood method to construct a confidence interval for a high quantile by using a smaller sample fraction $k$ to ensure that the involved high quantile estimator has a null asymptotic bias. The following theorem can be used to evaluate the coverage probability of the normal approximation based intervals via the high quantile estimator $\hat{x}_p$ in (2.56).

**Theorem 2.20.** *Under conditions of Theorem 2.16 with $\lambda = 0$,*

$$
\begin{aligned}
& P\Big(\frac{\hat{\alpha}(k)\sqrt{k}}{\log(np/k)} \log \frac{\hat{x}_p}{x_p} \le x\Big) - \Phi(x) \\
= & \frac{\phi(x)(1+2x^2)}{3\sqrt{k}} - \frac{\phi(x)\sqrt{k}A(k/n)}{\alpha(1+\rho)} - \frac{1}{2}x\phi(x)(\log\frac{k}{np})^{-2} \\
& + o\Big((\log\frac{k}{np})^{-2} + \frac{1}{\sqrt{k}} + \sqrt{k}|A(\frac{k}{n})|\Big),
\end{aligned}
$$

*where $\phi(x)$ and $\Phi(x)$ denote the density function and distribution function of a standard normal random variable.*

*Proof.* See Peng and Qi [83]. □

Next we introduce the following likelihood ratio method in Peng and Qi [83]. As before, a censored likelihood function is written as

$$
L(\alpha, c) = \prod_{i=1}^{n} \{c\alpha X_i^{-\alpha-1}\}^{\delta_i} \left\{1 - cX_{n,n-k}^{-\alpha}\right\}^{1-\delta_i} \quad \text{with} \quad \delta_i = I(X_i > X_{n,n-k}).
$$

First we obtain $l_1 = \max_{\alpha>0,c>0} \log L(\alpha, c)$. Second we maximize $\log L(\alpha, c)$ subject to

$$
\alpha > 0, \quad c > 0, \quad \alpha \log x_p + \log(p/c) = 0,
$$

and denote this maximized log-likelihood function as $l_2(x_p)$. The following Wilks theorem can be used to construct a confidence interval for the high quantile $x_p = \bar{F}^{-}(p)$.

**Theorem 2.21.** *Under conditions of Theorem 2.16 with $\lambda = 0$, we have*

$$
-2\left\{l_2(x_{p,0}) - l_1\right\} \xrightarrow{d} \chi^2(1) \quad \textit{as} \quad n \to \infty,
$$

*where $x_{p,0}$ denotes the true value of the high quantile $x_p = \bar{F}^{-}(p)$.*

*Proof.* See Peng and Qi [83]. □

Finally we introduce a profile empirical likelihood method, which is different from the data tilting method in Peng and Qi [83]. Write $\theta = \bar{F}^{-}(p)$, where $p = p_n \to 0$ as $n \to \infty$. Since $\theta$ can be expressed as $\theta = (p/c)^{-1/\alpha}$ under the ideal model $\bar{F}(x) = cx^{-\alpha}$, we have $c = p\theta^{\alpha}$ and then rewrite the censored likelihood function as

$$\prod_{i=1}^{n} \{p\theta^{\alpha}\alpha X_i^{-\alpha-1}\}^{\delta_i} \left\{1 - p\theta^{\alpha} X_{n,n-k}^{-\alpha}\right\}^{1-\delta_i} \quad \text{with } \delta_i = I(X_i > X_{n,n-k}),$$

which results in the score equations with respect to $\theta$ and $\alpha$ as

$$\sum_{i=1}^{n} \left\{ \frac{\delta_i \alpha}{\theta} - \frac{(1-\delta_i)p\alpha\theta^{\alpha-1}X_{n,n-k}^{-\alpha}}{1 - p\theta^{\alpha}X_{n,n-k}^{-\alpha}} \right\} = 0$$

and

$$\sum_{i=1}^{n} \left\{ \delta_i(\log\theta + \frac{1}{\alpha} - \log X_i) + \frac{(1-\delta_i)p\theta^{\alpha}X_{n,n-k}^{-\alpha}\big(\log(X_{n,n-k}) - \log(\theta)\big)}{1 - p\theta^{\alpha}X_{n,n-k}^{-\alpha}} \right\} = 0.$$

These are equivalent to

$$\sum_{i=1}^{n}(\delta_i - p\theta^{\alpha}X_{n,n-k}^{-\alpha}) = 0 \quad \text{and} \quad \sum_{i=1}^{n}\delta_i(1 - \alpha\log\frac{X_i}{X_{n,n-k}}) = 0.$$

Hence the empirical likelihood function for $(\alpha, \theta)^T$ is

$$L(\alpha,\theta) = \sup\left\{\prod_{i=1}^{n}(np_i) : p_1 \geq 0, \cdots, p_n \geq 0, \sum_{i=1}^{n} p_i = 1, \sum_{i=1}^{n} p_i \mathbf{Y}_i(\alpha,\theta) = 0\right\},$$

where $\mathbf{Y}_i(\alpha,\theta) = \big(Y_{i,1}(\alpha,\theta), Y_{i,2}(\alpha,\theta)\big)^T$,

$$Y_{i,1}(\alpha,\theta) = \delta_i - p\theta^{\alpha}X_{n,n-k}^{-\alpha} \text{ and } Y_{i,2}(\alpha,\theta) = \delta_i(1 - \alpha\log\frac{X_i}{X_{n,n-k}}).$$

By the Lagrange multiplier technique, we have $p_i = \frac{1}{n\left(1+\boldsymbol{\lambda}^T\mathbf{Y}_i(\alpha,\theta)\right)}$ and

$$l(\alpha,\theta) := -2\log L(\alpha,\theta) = 2\sum_{i=1}^{n}\log\big(1 + \boldsymbol{\lambda}^T\mathbf{Y}_i(\alpha,\theta)\big),$$

where $\boldsymbol{\lambda} = \boldsymbol{\lambda}(\alpha, \theta)$ satisfies

$$\sum_{i=1}^{n} \frac{\mathbf{Y}_i(\alpha, \theta)}{1 + \boldsymbol{\lambda}^T \mathbf{Y}_i(\alpha, \theta)} = 0.$$

Since we are interested in $\theta$, we consider the following profile empirical likelihood function

$$l^P(\theta) = \min_{\alpha>0} l(\alpha, \theta).$$

**Theorem 2.22.** *Under conditions of (2.4), (2.28) and for some $\delta > 0$*

$$\frac{\sqrt{k}A(k/n)}{(k/n)^{1/\alpha}\bar{F}^-(k/n)} = o(n^{-\delta}), \ p = p_n \to 0, \ np/k \to 0, \ \log(np) = o(\sqrt{k}),$$

*$l^P(\theta_0)$ converges in distribution to a chi-squared limit with one degree of freedom as $n \to \infty$, where $\theta_0$ denotes the true value of $\theta$.*

Before proving the above theorem, we need two lemmas and reparameterize $\alpha = \beta/\log(np/k)$.

**Lemma 2.2.** *Under conditions of Theorem 2.22, we have as $n \to \infty$*

$$\frac{1}{\sqrt{k}}\sum_{i=1}^{n} \mathbf{Y}_i(\alpha_0, \theta_0) \xrightarrow{d} N(0, I_{2\times 2}) \ \ \textit{and} \ \frac{1}{k}\sum_{i=1}^{n} \mathbf{Y}_i^T(\alpha_0, \theta_0)\mathbf{Y}_i(\alpha_0, \theta_0) \xrightarrow{p} I_{2\times 2},$$

*where $I_{2\times 2}$ denotes the $2 \times 2$ identity matrix.*

*Proof.* Write $X_{n,n-i+1} = \bar{F}^-(U_{n,i})$ and use (2.9) and (2.30), we can show that as $n \to \infty$

$$\begin{aligned}
\frac{1}{\sqrt{k}}\sum_{i=1}^{n} Y_{i,1}(\alpha_0, \theta_0) &= k^{1/2} - k^{-1/2}np\left(\frac{\bar{F}^-(U_{n,k+1})}{\bar{F}^-(p)}\right)^{-\alpha_0} \\
&= \sqrt{k}\left(\frac{n}{k}U_{n,k+1} - 1\right) + o_p(1) \\
&= W_n(1) + o_p(1)
\end{aligned}$$

with $W_n(s)$ given in (2.30), and

$$\begin{aligned}
\frac{1}{k}\sum_{i=1}^{n} Y_{i,1}^2(\alpha_0, \theta_0) &= 1 - 2p\theta_0^{\alpha_0}X_{n,n-k}^{-\alpha_0} + k^{-1}np^2\theta_0^{2\alpha_0}X_{n,n-k}^{-2\alpha_0} \\
&= 1 - 2\frac{k}{n} + \frac{k}{n} + o_p(1) \\
&= 1 + o_p(1).
\end{aligned}$$

It follows from (2.9), (2.30) and the same arguments in the proof of Theorem 2.4 that

$$\frac{1}{\sqrt{k}}\sum_{i=1}^{n} Y_{i,2}(\alpha_0,\theta_0) = -\int_0^1 \left\{\frac{W_n(s)}{s} - W_n(1)\right\} ds + o_p(1)$$

and

$$\frac{1}{k}\sum_{i=1}^{n} Y_{i,2}^2(\alpha_0,\theta_0) = 1 + o_p(1) \quad \text{as} \quad n \to \infty.$$

Hence the lemma follows from the above equations. □

**Lemma 2.3.** *Under conditions of Theorem 2.22, $l(\beta,\theta_0)$ attains its minimum value with probability one at some point $\bar{\beta}$ such that $|\bar{\beta}-\beta_0| < k^{-1/3}$, and $\bar{\beta}$ and $\bar{\boldsymbol{\lambda}}$ satisfy*

$$Q_{1n}(\bar{\beta},\bar{\boldsymbol{\lambda}}) = 0 \text{ and } Q_{2n}(\bar{\beta},\bar{\boldsymbol{\lambda}}) = 0,$$

*where*

$$Q_{1n}(\beta,\boldsymbol{\lambda}) = \frac{1}{k}\sum_{i=1}^{n} \frac{\mathbf{Y}_i(\beta,\theta_0)}{1+\boldsymbol{\lambda}^T\mathbf{Y}_i(\beta,\theta_0)}$$

*and*

$$Q_{2n}(\beta,\boldsymbol{\lambda}) = \frac{1}{k}\sum_{i=1}^{n} \frac{1}{1+\boldsymbol{\lambda}^T\mathbf{Y}_i(\beta,\theta_0)}\left\{\frac{d}{d\beta}\mathbf{Y}_i^T(\beta,\theta_0)\right\}\boldsymbol{\lambda}.$$

*Proof.* Recall $\beta = \alpha\log(np/k)$. First we have

$$\frac{1}{k}\sum_{i=1}^{n} \frac{dY_{i,1}(\beta_0,\theta_0)}{d\beta} \xrightarrow{p} \alpha_0^{-1}, \quad \frac{1}{k}\sum_{i=1}^{n} \frac{dY_{i,2}(\beta_0,\theta_0)}{d\beta} \xrightarrow{p} 0,$$

and

$$\begin{aligned}\boldsymbol{\lambda}(\beta) &= \left\{\frac{1}{k}\sum_{i=1}^{n}\mathbf{Y}_i(\beta,\theta_0)\mathbf{Y}_i^T(\beta,\theta_0)\right\}^{-1}\frac{1}{k}\sum_{i=1}^{n}\mathbf{Y}_i(\beta,\theta_0) + o_p(k^{-1/3})\\ &= o_p(k^{-1/3})\end{aligned}$$

uniformly for $|\beta-\beta_0| \le k^{-1/3}$, which implies that for $|\beta-\beta_0| = k^{-1/3}$,

$$
\begin{aligned}
&l(\beta, \theta_0) \\
&= 2\sum_{i=1}^{n} \boldsymbol{\lambda}^T \mathbf{Y}_i(\beta, \theta_0) - \sum_{i=1}^{n} \{\boldsymbol{\lambda}^T \mathbf{Y}_i(\beta, \theta_0)\}^2 + o_p(k^{1/3}) \\
&= k\left\{\frac{1}{k}\sum_{i=1}^{n} \mathbf{Y}_i(\beta, \theta_0)\right\}^T \left\{\frac{1}{k}\sum_{i=1}^{n} \mathbf{Y}_i(\beta, \theta_0)\mathbf{Y}_i^T(\beta, \theta_0)\right\}^{-1} \left\{\frac{1}{k}\sum_{i=1}^{n} \mathbf{Y}_i(\beta, \theta_0)\right\} \\
&\quad + o_p(k^{1/3}) \\
&= k\left\{\frac{1}{k}\sum_{i=1}^{n} \mathbf{Y}_i(\beta_0, \theta_0) + \frac{1}{k}\sum_{i=1}^{n} \frac{d\mathbf{Y}_i(\beta_0, \theta_0)}{d\beta}(\beta - \beta_0)\right\}^T \\
&\quad \times \left\{\frac{1}{k}\sum_{i=1}^{n} \mathbf{Y}_i(\beta_0, \theta_0)\mathbf{Y}_i^T(\beta_0, \theta_0)\right\}^{-1} \\
&\quad \times \left\{\frac{1}{k}\sum_{i=1}^{n} \mathbf{Y}_i(\beta_0, \theta_0) + \frac{1}{k}\sum_{i=1}^{n} \frac{d\mathbf{Y}_i(\beta_0, \theta_0)}{d\beta}(\beta - \beta_0)\right\} + o_p(k^{-1/3}) \\
&= k^{1/3}\alpha_0^{-2} + o_p(k^{1/3}).
\end{aligned}
$$

Similarly, we have $l(\beta_0, \theta_0) = o_p(1)$. Hence the lemma follows. □

*Proof of Theorem 2.22.* First we have

$$
\frac{\partial Q_{1n}(\beta_0, 0)}{\partial \beta} \xrightarrow{p} (\alpha_0^{-1}, 0)^T, \quad \frac{\partial Q_{1n}(\beta_0, 0)}{\partial \boldsymbol{\lambda}} \xrightarrow{p} -I_{2\times 2},
$$
$$
\frac{\partial Q_{2n}(\beta_0, 0)}{\partial \beta} = 0, \quad \frac{\partial Q_{2n}(\beta_0, 0)}{\partial \boldsymbol{\lambda}} \xrightarrow{p} (\alpha_0^{-1}, 0)^T.
$$

By Lemma 2.3 and Taylor expansions, we have

$$
0 = Q_{1n}(\bar{\beta}, \bar{\boldsymbol{\lambda}}) = Q_{1n}(\beta_0, 0) + \frac{dQ_{1n}(\beta_0, 0)}{d\beta}(\bar{\beta} - \beta_0) + \frac{\partial Q_{1n}(\beta_0, 0)}{\partial \boldsymbol{\lambda}^T}\bar{\boldsymbol{\lambda}} + o_p(\delta_n)
$$

and

$$
0 = Q_{2n}(\bar{\beta}, \bar{\boldsymbol{\lambda}}) = Q_{2n}(\beta_0, 0) + \frac{dQ_{2n}(\beta_0, 0)}{d\beta}(\bar{\beta} - \beta_0) + \frac{\partial Q_{2n}(\beta_0, 0)}{\partial \boldsymbol{\lambda}^T}\bar{\boldsymbol{\lambda}} + O_p(\delta_n^2),
$$

where $\delta_n = |\bar{\beta} - \beta_0| + ||\bar{\boldsymbol{\lambda}}||$, which imply that

$$
\begin{pmatrix} \bar{\boldsymbol{\lambda}} \\ \bar{\beta} - \beta_0 \end{pmatrix} = S_n^{-1} \begin{pmatrix} -Q_{1n}(\alpha_0, 0) + o_p(\delta_n) \\ o_p(\delta_n) \end{pmatrix},
$$

where

$$S_n = \begin{pmatrix} \dfrac{\partial Q_{1n}(\beta_0, 0)}{\partial \boldsymbol{\lambda}^T} & \dfrac{dQ_{1n}(\beta_0, 0)}{d\beta} \\ \dfrac{\partial Q_{2n}(\beta_0, 0)}{\partial \boldsymbol{\lambda}^T} & 0 \end{pmatrix} \xrightarrow{p} \begin{pmatrix} -I_{2\times 2} & S_{12} \\ S_{21} & 0 \end{pmatrix}$$

with $S_{12} = S_{21}^T = (\alpha_0^{-1}, 0)^T$. It follows from Lemma 2.2 and the above expansions that

$$\begin{aligned} l^P(\theta_0) &= 2k(\bar{\boldsymbol{\lambda}}^T, \bar{\beta} - \beta_0)\big(Q_{1n}^T(\beta_0, 0), 0\big)^T + k(\bar{\boldsymbol{\lambda}}^T, \bar{\beta} - \beta_0) S_n (\bar{\boldsymbol{\lambda}}, \bar{\beta} - \beta_0)^T \\ &\quad + o_p(1) \\ &= -k\big(Q_{1n}^T(\beta_0, 0), 0\big) S_n^{-1} \big(Q_{1n}(\beta_0, 0), 0\big)^T + o_p(1) \\ &= -(\xi^T, 0) \begin{pmatrix} -I_{2\times 2} & S_{12} \\ S_{12}^T & 0 \end{pmatrix}^{-1} (\xi^T, 0)^T + o_p(1), \end{aligned}$$

where $\xi \sim N(0, I_{2\times 2})$. Since

$$\begin{pmatrix} -I_{2\times 2} & S_{12} \\ S_{12}^T & 0 \end{pmatrix}^{-1} = \begin{pmatrix} -I_{2\times 2} + \alpha_0^2 S_{12} S_{12}^T & \alpha_0^2 S_{12} \\ \alpha_0^2 S_{12}^T & \alpha_0^2 \end{pmatrix},$$

we have

$$-(\xi^T, 0) \begin{pmatrix} -I_{2\times 2} & S_{12} \\ S_{12}^T & 0 \end{pmatrix}^{-1} (\xi^T, 0)^T = \xi^T \begin{pmatrix} 0 & 0 \\ 0 & 1 \end{pmatrix} \xi \sim \chi^2(1),$$

i.e., the theorem follows. □

## 2.6 GOODNESS-OF-FIT TESTS

Heavy tailed distributions are often used to model losses in insurance and finance, and predicting an extreme event based on modeling the tail is robust since it only assumes some tail behavior of the underlying distribution without fitting a parametric family to the whole distribution. On the other hand, one sacrifices the efficiency if an inference only uses a limited observations in the tail region. Here we introduce some methods for testing whether a distribution function has a heavy tail, that is, whether a distribution function satisfies (2.1) based on a random sample $X_1, \cdots, X_n$.

Define

$$G_k(s) = \frac{1}{k}\sum_{i=1}^{k} I(\frac{X_{n,n-i+1}}{X_{n,n-k}} \le s) \quad \text{and} \quad G(s;\alpha) = 1 - s^{-\alpha} \quad \text{for} \quad s \ge 1,$$

where $X_{n,1} \le \cdots \le X_{n,n}$ denote the order statistics of $X_1, \cdots, X_n$. The following seven test statistics can be employed to test condition (2.1), where $\hat{\alpha}(k)$ is the Hill estimator in (2.31).

- *Kolmogorov–Smirnov test:* $\sup_{s\ge 1} |KS(s;\hat{\alpha}(k))|$, where

$$KS(s;\alpha) = 1 - G_k(s) - s^{-\alpha}.$$

- *Berk–Jones test:* $\sup_{s\ge 1}\left\{kBJ(s;\hat{\alpha}(k))\right\}$, where

$$BJ(s;\alpha) = 2K(G_k(s), 1 - s^{-\alpha})$$

  with

$$K(p_1, p_2) = p_1 \log\frac{p_1}{p_2} + (1-p_1)\log\frac{1-p_1}{1-p_2}.$$

- *Estimated score test:* $\sup_{s\ge 1} |SC(s;\hat{\alpha}(k))|$, where

$$SC(s;\alpha) = G_k(s) - \alpha\int_1^s \frac{1-G_k(t)}{t}\,dt.$$

- *Cramér–von Mises test:* $KSI = \int_1^\infty \left\{\sqrt{k}KS(s;\hat{\alpha}(k))\right\}^2 dG(s;\hat{\alpha}(k))$.
- *Integrated Berk–Jones test:* $BJI = \int_1^\infty kBJ(s;\hat{\alpha}(k))\,dG(s;\hat{\alpha}(k))$.
- *Integrated score test:* $SCI = \int_1^\infty \left\{\sqrt{k}SC(s;\hat{\alpha}(k))\right\}^2 dG(s;\hat{\alpha}(k))$.
- *Jackson test:* $JT = \frac{\sum_{i=1}^k c_{k-i+1,k} i\log(X_{n,n-i+1}/X_{n,n-i})}{\sum_{i=1}^k i\log(X_{n,n-i+1}/X_{n,n-i})}$, where $c_{k-i+1,k} = 1 - \log\frac{i+1}{k+1}$.

Koning and Peng [65] discussed the first six test statistics, derived their asymptotic distributions and studied the Bahadur efficiency. Beirlant et al. [6] studied the last test statistic by extending the Jackson test for exponential distributions in Jackson [62] to heavy tails.

**Theorem 2.23.** *Under conditions (2.4), (2.28) and*

$$\sqrt{k}\frac{A(k/n)}{(k/n)^{1/\alpha}\bar{F}^{-}(k/n)} \to 0 \quad as \quad n \to \infty,$$

*we have as* $n \to \infty$

$$\begin{aligned}&\big(\sup_{s\geq 1}|\sqrt{k}KS(s;\hat{\alpha}(k))|, \sup_{s\geq 1}|\sqrt{k}SC(s;\hat{\alpha}(k))|\big)\\ &\xrightarrow{d} \Big(\sup_{0<v<1}|B(v)+v\log(v)\int_0^1 s^{-1}B(s)\,ds|,\ \sup_{0<v<1}|\bar{B}(v)|\Big),\end{aligned} \tag{2.63}$$

$$(KSI, BJI, SCI) \xrightarrow{d} \Big(\int_0^1 W^2(v)\,dv, \int_0^1 \frac{W^2(v)}{v(1-v)}\,dv, \int_0^1 \bar{B}^2(v)\,dv\Big) \tag{2.64}$$

*and*

$$\sqrt{k}\{JT-2\} \xrightarrow{d} N(0,1), \tag{2.65}$$

*where* $B(v)$ *and* $\bar{B}(v) = B(v) - \int_0^v s^{-1}B(s)\,ds + v\int_0^1 s^{-1}B(s)\,ds$ *are Brownian bridges, and* $W(v) = B(v) + v\log(v)\int_0^1 s^{-1}B(s)\,ds$ *is a Gaussian process.*

*Proof.* Note that for $r > 1$

$$\begin{aligned}1-G_k(r) &= 1-\frac{1}{k}\sum_{i=1}^{k} I\Big(\frac{X_{n,n-k+i}}{X_{n,n-k}} \leq r\Big) = \frac{1}{k}\sum_{i=1}^{k} I\Big(\frac{X_{n,n-k+i}}{X_{n,n-k}} > r\Big)\\ &= \frac{1}{k}\sum_{i=1}^{n} I\Big(\frac{X_i}{X_{n,n-k}} > r\Big) = \frac{1}{k}\sum_{i=1}^{n} I(U_i < \bar{F}(rX_{n,n-k}))\end{aligned} \tag{2.66}$$

by defining $\bar{F}(X_i) = U_i$ for $1 \leq i \leq n$. Further write

$$\begin{aligned}1-G_k(r)-r^{-\alpha} &= \frac{1}{k}\sum_{i=1}^{n} I\big(U_i < \frac{k}{n}\frac{n}{k}\bar{F}(rX_{n,n-k})\big) - \frac{n}{k}\bar{F}(rX_{n,n-k})\\ &\quad + \Big(\frac{\bar{F}(rX_{n,n-k})}{\bar{F}(X_{n,n-k})} - r^{-\alpha}\Big)\frac{n}{k}\bar{F}(X_{n,n-k}) + r^{-\alpha}\Big(\frac{n}{k}\bar{F}(X_{n,n-k}) - 1\Big)\end{aligned}$$

for $r > 1$. Therefore by noting that

$$\frac{\bar{F}(rX_{n,n-k})}{\bar{F}(X_{n,n-k})} \xrightarrow{p} r^{-\alpha} \quad \text{and} \quad \frac{n}{k}\bar{F}(X_{n,n-k}) \xrightarrow{p} 1 \quad \text{as} \quad n \to \infty,$$

and using (2.29) and (2.30), we have

$$\sup_{r>1}|\sqrt{k}\{1-G_k(r)-r^{-\alpha}\}-\{W_n(r^{-\alpha})-r^{-\alpha}W_n(1)\}|\xrightarrow{p}0 \quad \text{as} \quad n\to\infty, \tag{2.67}$$

where $W_n(s)$ is given in (2.30). From the proof of Theorem 2.4, we have

$$\sqrt{k}\{\hat{\alpha}(k)-\alpha\}=-\alpha\int_0^1\frac{W_n(s)-sW_n(1)}{s}\,ds+o_p(1) \quad \text{as} \quad n\to\infty. \tag{2.68}$$

Write

$$\begin{aligned}&\sup_{r>1}|\sqrt{k}KS(r;\hat{\alpha}(k))|\\&=\sup_{r>1}|\sqrt{k}\{(1-G_k(r))-r^{-\alpha}\}+\sqrt{k}\left\{r^{-\alpha}-r^{-\hat{\alpha}(k)}\right\}|.\end{aligned}$$

It follows from (2.67) and (2.68) that

$$\begin{aligned}&\sup_{r>1}|\sqrt{k}KS(r;\hat{\alpha}(k))|\\&\xrightarrow{d}\sup_{r>1}|W_n(r^{-\alpha})-r^{-\alpha}W_n(1)-r^{-\alpha}\log(r)\alpha\int_0^1\frac{W_n(s)-sW_n(1)}{s}\,ds| \qquad (2.69)\\&=\sup_{0<v<1}|W_n(v)-vW_n(1)+v\log(v)\int_0^1\frac{W_n(s)-sW_n(1)}{s}\,ds|.\end{aligned}$$

Write

$$\begin{aligned}&\sup_{r>1}|\sqrt{k}SC(r;\hat{\alpha}(k))|\\&=\sup_{r>1}|\sqrt{k}G_k(r)-\sqrt{k}\hat{\alpha}(k)\int_1^r\frac{1-G_k(s)}{s}\,ds|\\&=\sup_{r>1}|\sqrt{k}G_k(r)-\sqrt{k}\alpha\int_0^r\frac{1-G_k(s)}{s}\,ds\\&\quad-\sqrt{k}(\hat{\alpha}(k)-\alpha)\int_1^r\frac{1-G_k(s)}{s}\,ds|\\&=\sup_{r>1}|\sqrt{k}\{G_k(r)-1+r^{-\alpha}\}-\sqrt{k}\alpha\int_1^r\frac{1-G_k(s)-s^{-\alpha}}{s}\,ds\\&\quad-\sqrt{k}\{\hat{\alpha}(k)-\alpha\}\frac{1-r^{-\alpha}}{\alpha}-\sqrt{k}\{\hat{\alpha}(k)-\alpha\}\int_1^r\frac{1-G_k(s)-s^{-\alpha}}{s}\,ds|,\end{aligned}$$

which implies that

$$\begin{aligned}
&\sup_{r>1} |\sqrt{k} SC(r;\hat{\alpha}(k))| \\
&\xrightarrow{d} \sup_{r>1} \Big| -\left\{W_n(r^{-\alpha}) - r^{-\alpha}W_n(1)\right\} - \alpha\int_1^r \frac{W_n(s^{-\alpha}) - s^{-\alpha}W_n(1)}{s}\,ds \\
&\quad + (1-r^{-\alpha})\int_0^1 \frac{W_n(s) - sW_n(1)}{s}\,ds\Big| \\
&= \sup_{0<v<1} \Big| -\{W_n(v) - vW_n(1)\} - \int_v^1 \frac{W_n(s) - sW_n(1)}{s}\,ds \\
&\quad + (1-v)\int_0^1 \frac{W_n(s) - sW_n(1)}{s}\,ds\Big| \\
&= \sup_{0<v<1} \Big| -\{W_n(v) - vW_n(1)\} + \int_0^v \frac{W_n(s) - sW_n(1)}{s}\,ds \\
&\quad - v\int_0^1 \frac{W_n(s) - sW_n(1)}{s}\,ds\Big|
\end{aligned} \tag{2.70}$$

by using (2.67) and (2.68). Hence (2.63) follows from (2.69), (2.70) and

$$\{W_n(s) - sW_n(1) : 0 < s < 1\} \stackrel{d}{=} \{B(s) : 0 < s < 1\}. \tag{2.71}$$

For proving (2.64), we only need to derive the convergence of BJI since the convergence of the other two terms follows from (2.69) and (2.70).

For $\epsilon \in (0, 1/2)$, write

$$\begin{aligned}
&\int_1^\infty BJ(r;\hat{\alpha}(k))\,dG(r;\hat{\alpha}(k)) \\
&= \int_0^1 BJ(v^{-1/\hat{\alpha}(k)};\hat{\alpha}(k))\,dv \\
&= \int_0^\epsilon BJ(v^{-1/\hat{\alpha}(k)};\hat{\alpha}(k))\,dv + \int_\epsilon^{1-\epsilon} BJ(v^{-1/\hat{\alpha}(k)};\hat{\alpha}(k))\,dv \\
&\quad + \int_{1-\epsilon}^1 BJ(v^{-1/\hat{\alpha}(k)};\hat{\alpha}(k))\,dv.
\end{aligned} \tag{2.72}$$

By (2.67), (2.68) and Taylor expansions, we have

$$\begin{aligned}
&k\int_\epsilon^{1-\epsilon} BJ(v^{-1/\hat{\alpha}(k)};\hat{\alpha}(k))\,dv \\
&= k\int_\epsilon^{1-\epsilon} \Big\{2G_k(v^{-1/\hat{\alpha}(k)})\log\frac{G_k(v^{-1/\hat{\alpha}(k)})}{1-v}
\end{aligned}$$

$$
\begin{aligned}
&+2\big(1-G_k(v^{-1/\hat\alpha(k)})\big)\log\frac{1-G_k(v^{-1/\hat\alpha(k)})}{v}\Big\}\,dv\\
&=k\int_\epsilon^{1-\epsilon}\Big\{2G_k(v^{-1/\hat\alpha(k)})\Big(\frac{G_k(v^{-1/\hat\alpha(k)})}{1-v}-1\Big)\\
&-G_k(v^{-1/\hat\alpha(k)})\Big(\frac{G_k(v^{-1/\hat\alpha(k)})}{1-v}-1\Big)^2\\
&+2\big(1-G_k(v^{-1/\hat\alpha(k)})\big)\Big(\frac{1-G_k(v^{-1/\hat\alpha(k)})}{v}-1\Big)\\
&-\big(1-G_k(v^{-1/\hat\alpha(k)})\big)\Big(\frac{1-G_k(v^{-1/\hat\alpha(k)})}{v}-1\Big)^2\Big\}\,dv+o_p(1)\\
&=k\int_\epsilon^{1-\epsilon}\Big\{2\frac{(G_k(v^{-1/\hat\alpha(k)})-1+v)^2}{1-v}+2\big(G_k(v^{-1/\hat\alpha(k)})-1+v\big) \qquad (2.73)\\
&-\frac{(G_k(v^{-1/\hat\alpha(k)})-1+v)^3}{(1-v)^2}-\frac{(G_k(v^{-1/\hat\alpha(k)}-1+v)^2}{1-v}\\
&+2\frac{(1-G_k(v^{-1/\hat\alpha(k)})-v)^2}{v}+2\big(1-G_k(v^{-1/\hat\alpha(k)})-v\big)\\
&-\frac{(1-G_k(v^{-1/\hat\alpha(k)})-v)^3}{v^2}-\frac{(1-G_k(v^{-1/\hat\alpha(k)})-v)^2}{v}\Big\}\,dv+o_p(1)\\
&=k\int_\epsilon^{1-\epsilon}\Big\{\frac{(G_k(v^{-1/\hat\alpha(k)})-1+v)^2}{1-v}+\frac{(1-G_k(v^{-1/\hat\alpha(k)})-v)^2}{v}\Big\}\,dv+o_p(1)\\
&=k\int_\epsilon^{1-\epsilon}\frac{(1-G_k(v^{-1/\hat\alpha(k)})-v)^2}{v(1-v)}\,dv+o_p(1)\\
&\xrightarrow{d}\int_\epsilon^{1-\epsilon}\Big\{v(1-v)\Big\}^{-1}\Big\{W_n(v)-vW_n(1)\\
&+v\log(v)\int_0^1\frac{W_n(s)-sW_n(1)}{s}\,ds\Big\}^2\,dv \quad\text{as}\quad n\to\infty.
\end{aligned}
$$

For $0<s,t<1$, write

$$
\begin{aligned}
&K(s,1-t)\\
&=-s\Big\{\log(1+\frac{1-t-s}{2s})-\frac{1-t-s}{2s}\Big\}+s\Big\{\log(1-\frac{1-t-s}{2(1-t)})+\frac{1-t-s}{2(1-t)}\Big\}\\
&-(1-s)\Big\{\log(1-\frac{1-t-s}{2(1-s)})+\frac{1-t-s}{2(1-s)}\Big\}\\
&+(1-s)\Big\{\log(1+\frac{1-t-s}{2t})-\frac{1-t-s}{2t}\Big\}+\frac{(1-t-s)^2}{2t(1-t)}.
\end{aligned}
$$

Since

$$|\log(1+y)-y| \le y^2 \quad \text{for} \quad y \ge -\frac{1}{2},$$

we have

$$\begin{aligned}
&\frac{1}{2}BJ\big(v^{-1/\hat{\alpha}(k)};\hat{\alpha}(k)\big) = K(G_k(v^{-1/\hat{\alpha}(k)}),1-v)\\
&\le \frac{\{1-v-G_k(v^{-1/\hat{\alpha}(k)})\}^2}{4G_k(v^{-1/\hat{\alpha}(k)})} + G_k(v^{-1/\hat{\alpha}(k)})\frac{\{1-v-G_k(v^{-1/\hat{\alpha}(k)})\}^2}{4(1-v)^2}\\
&\quad + \frac{\{1-v-G_k(v^{-1/\hat{\alpha}(k)})\}^2}{4\{1-G_k(v^{-1/\hat{\alpha}(k)})\}} + \Big\{1-G_k(v^{-1/\hat{\alpha}(k)})\Big\}\frac{\{1-v-G_k(v^{-1/\hat{\alpha}(k)})\}^2}{4v^2}\\
&\quad + \frac{\{1-v-G_k(v^{-1/\hat{\alpha}(k)})\}^2}{2v(1-v)}
\end{aligned} \tag{2.74}$$

when $0 < G_k(v^{-1/\hat{\alpha}(k)}) < 1$, and

$$\frac{1}{2}BJ\big(v^{-1/\hat{\alpha}(k)});\hat{\alpha}(k)\big) = \begin{cases} -\log(v) & \text{if} \quad G_k(v^{-1/\hat{\alpha}(k)}) = 0,\\ -\log(1-v) & \text{if} \quad G_k(v^{-1/\hat{\alpha}(k)}) = 1.\end{cases}$$

It follows from the definition of $G_k(r)$ that $G_k(v^{-1/\hat{\alpha}(k)}) = 0$ is equivalent to

$$\frac{X_{n,n-k+1}}{X_{n,n-k}} > v^{-1/\hat{\alpha}(k)}, \quad \text{i.e.,} \quad v > (\frac{X_{n,n-k+1}}{X_{n,n-k}})^{-\hat{\alpha}(k)},$$

which implies

$$\begin{aligned}
&k\int_0^{\epsilon} BJ(v^{-1/\hat{\alpha}(k)};\hat{\alpha}(k))I\big(G_k(v^{-1/\hat{\alpha}(k)})=0\big)\,dv\\
&= -2k\int_0^{\epsilon}\log(v)I\big(v > (\frac{X_{n,n-k+1}}{X_{n,n-k}})^{-\hat{\alpha}(k)}\big)\,dv\\
&= o_p(1)
\end{aligned} \tag{2.75}$$

by using $(X_{n,n-k+1}/X_{n,n-k})^{-\hat{\alpha}(k)} \xrightarrow{p} 1$, and

$$\begin{aligned}
&k\int_{1-\epsilon}^{1} BJ(v^{-1/\hat{\alpha}(k)};\hat{\alpha}(k))I\big(G_k(v^{-1/\hat{\alpha}(k)})=0\big)\,dv\\
&= -2k\int_{1-\epsilon}^{1}\log(v)I\big(v > (\frac{X_{n,n-k+1}}{X_{n,n-k}})^{-\hat{\alpha}(k)}\big)\,dv\\
&= O_p\big(-2k\log((\frac{X_{n,n-k+1}}{X_{n,n-k}})^{-\hat{\alpha}(k)})\Big\{1-(\frac{X_{n,n-k+1}}{X_{n,n-k}})^{-\hat{\alpha}(k)}\Big\}\big).
\end{aligned} \tag{2.76}$$

By (2.9), (2.30) and Theorem 2.4, we have

$$\begin{aligned}&\sqrt{k}\left\{1-(\frac{X_{n,n-k+1}}{X_{n,n-k}})^{-\hat{\alpha}(k)}\right\}=\sqrt{k}\left\{1-(\frac{X_{n,n-k+1}}{X_{n,n-k}})^{-\alpha}\right\}+o_p(1)\\&=\sqrt{k}\left\{1-\frac{U_{n,k-1}}{U_{n,k}}\right\}+o_p(1)=o_p(1).\end{aligned}\tag{2.77}$$

Therefore it follows from (2.76) and (2.77) that

$$k\int_{1-\epsilon}^{1}BJ\left(v^{-1/\hat{\alpha}(k)};\hat{\alpha}(k)\right)I\left(G_k(v^{-1/\hat{\alpha}(k)})=0\right)dv=o_p(1).\tag{2.78}$$

Similarly

$$k\int_{0}^{\epsilon}BJ(v^{-1/\hat{\alpha}(k)};\hat{\alpha}(k))I\left(G_k(v^{-1/\hat{\alpha}(k)})=1\right)dv=o_p(1)\tag{2.79}$$

and

$$k\int_{1-\epsilon}^{1}BJ(v^{-1/\hat{\alpha}(k)};\hat{\alpha}(k))I\left(G_k(v^{-1/\hat{\alpha}(k)})=1\right)dv=o_p(1).\tag{2.80}$$

Define set $A=\left\{v:0<G_k(v^{-1/\hat{\alpha}(k)})<1\right\}$. By (2.66) and (2.24), we have

$$\sup_{v\in A}\frac{\frac{n}{k}\bar{F}(v^{-1/\hat{\alpha}(k)}X_{n,n-k})}{1-G_k(v^{-1/\hat{\alpha}(k)})}=O_p(1).\tag{2.81}$$

It follows from (2.29) that for any $\delta_1\in(0,1/2)$

$$\sup_{v\in A}\frac{|\sqrt{k}\left\{1-G_k(v^{-1/\hat{\alpha}(k)})-\bar{F}(v^{-1/\hat{\alpha}(k)}X_{n,n-k})\frac{n}{k}\right\}|}{\left\{\bar{F}(v^{-1/\hat{\alpha}(k)}X_{n,n-k})\frac{n}{k}\right\}^{\delta_1}}=O_p(1).\tag{2.82}$$

By (2.6) and Theorem 2.4, we have

$$\begin{aligned}\frac{|\sqrt{k}\left\{\bar{F}(v^{-1/\hat{\alpha}(k)}X_{n,n-k})\frac{n}{k}-v\right\}|}{v^{\delta_1}}&\le\frac{|\sqrt{k}\left\{\frac{\bar{F}(v^{-1/\hat{\alpha}(k)}X_{n,n-k})}{\bar{F}(X_{n,n-k})}-v^{\alpha/\hat{\alpha}(k)}\right\}|\frac{n}{k}\bar{F}(X_{n,n-k})}{v^{\delta_1}}\\&+\frac{|\sqrt{k}\left\{\frac{n}{k}\bar{F}(X_{n,n-k})-1\right\}|v^{\alpha/\hat{\alpha}(k)}}{v^{\delta_1}}+\frac{|\sqrt{k}\left\{v^{\alpha/\hat{\alpha}(k)}-v\right\}|}{v^{\delta_1}}=O_p(v^{-\delta_1-\delta_2})\end{aligned}\tag{2.83}$$

uniformly for $v \in A$ and any small $\delta_2 > 0$. Using (2.81)–(2.83) we can show that

$$\sup_{v\in A} v^{1-2\delta_3} \frac{k\left\{1 - v - G_k(v^{-1/\hat{\alpha}(k)})\right\}^2}{4(1 - G_k(v^{-1/\hat{\alpha}(k)}))} = O_p(1)$$

and

$$\sup_{v\in A} v^{2-3\delta_3} \left\{1 - G_k(v^{-1/\hat{\alpha}(k)})\right\} \frac{k\left\{1 - v - G_k(v^{-1/\hat{\alpha}(k)})\right\}^2}{4v^2} = O_p(1)$$

for any $\delta_3 \in (0, 1/2)$, which imply that

$$\begin{aligned} k\int_0^{\epsilon} BJ(v^{-1/\hat{\alpha}(k)}; \hat{\alpha}(k))I(A)\, dv &= O_p(\int_0^{\epsilon} v^{2\delta_3 - 1}\, dv) + O_p(\int_0^{\epsilon} v^{3\delta_3 - 2}\, dv) \\ &= O_p(1)\epsilon \end{aligned} \tag{2.84}$$

by using (2.74) and taking $\delta_3$ close to 1/2. Similarly

$$k\int_{1-\epsilon}^{1} BJ(v^{-1/\hat{\alpha}(k)}; \hat{\alpha}(k))I(A)\, dv = O_p(1)\epsilon. \tag{2.85}$$

Hence the convergence of IBJ in (2.64) follows from (2.72), (2.73), (2.75), (2.78), (2.79), (2.80), (2.84) and (2.85) by letting $\epsilon \to 0$.

Finally (2.65) can be shown in the same way as the proof of Theorem 2.11 or Theorem 2.4. □

**Remark 2.9.** The above proposed tests are distribution free in the sense that the asymptotic distributions are independent of the underlying distribution. Therefore, critical values can be tabulated. By simulating 100,000 random samples of Wiener processes on [0, 1] with 1000 equally spaced grid points, Koning and Peng [65] obtained the approximate critical values with level 0.95 for the above test statistics SC, SCI, KS, KSI and BJI in (2.63)–(2.65), which are 1.338, 0.456, 1.076, 0.220 and 1.313, respectively. Critical values for other levels can be found in Appendix A.

## 2.7 ESTIMATION OF MEAN

Let $X, X_1, \cdots, X_n$ be independent and identically distributed nonnegative random variables with distribution function $F$ satisfying (2.1). Suppose we are interested in estimating $\mu = E(X) < \infty$. The simple estimator, sample mean $\frac{1}{n}\sum_{i=1}^{n} X_i$, will have a nonnormal limit when $EX_i^2 = \infty$ (see Feller [37]). Here we propose to use (2.1) to get an estimator which always has a normal limit. More specifically the idea is to estimate the tail part parametrically and the middle part nonparametrically. The proposed method works for estimating $E(X^r)$ with any $r > 0$ as long as $E(X^r) < \infty$ and $X$ has a heavy tail. Moreover for $X$ taking values in $\mathbb{R}$, one can consider the positive part and negative part separately.

Throughout we use (1.2), (1.3) and (1.4) for concerning integrals, and write

$$E(X) = \int_0^\infty \bar{F}(x)\,dx = \int_0^{\bar{F}^-(k/n)} \bar{F}(x)\,dx + \int_{\bar{F}^-(k/n)}^\infty \bar{F}(x)\,dx =: \mu_1 + \mu_2.$$

By noting that

$$\mu_2 = \frac{k}{n}\int_{\bar{F}^-(k/n)}^\infty \frac{\bar{F}(x)}{\bar{F}(\bar{F}^-(k/n))}\,dx \sim \frac{k}{n}\int_{\bar{F}^-(k/n)}^\infty \left(\frac{x}{\bar{F}^-(k/n)}\right)^{-\alpha} dx = \frac{k}{n}\frac{\bar{F}^-(k/n)}{\alpha - 1},$$

we estimate $\mu_1$ and $\mu_2$, respectively, by

$$\hat{\mu}_1 = \int_0^{X_{n,n-k}} \bar{F}_n(x)\,dx \quad \text{and} \quad \hat{\mu}_2 = \frac{k}{n}\frac{X_{n,n-k}}{\hat{\alpha}(k) - 1},$$

where $\hat{\alpha}(k)$ is the Hill estimator defined in (2.31),

$$\bar{F}_n(x) = 1 - F_n(x) \quad \text{and} \quad F_n(x) = \frac{1}{n}\sum_{i=1}^n I(X_i \le x).$$

**Theorem 2.24.** *Under conditions of (2.4) with $\rho > 0$, (2.28) and*

$$\lim_{n\to\infty} \sqrt{k}\frac{A(k/n)}{(k/n)^{1/\alpha}\bar{F}^-(k/n)} = 0,$$

*we have*

$$\begin{aligned}&\frac{\sqrt{n}}{\sigma(k/n)}\left\{\hat{\mu}_1 + \hat{\mu}_2 - \mu\right\}\\ &\xrightarrow{d} N\Big(0, 1 + \Big(\frac{(2-\alpha)(2\alpha^2 - 2\alpha + 1)}{2(\alpha-1)^4} + \frac{2-\alpha}{\alpha-1}\Big)I(\alpha < 2)\Big)\end{aligned}$$

*as $n \to \infty$, where*

$$\sigma^2(s) = \int_0^{1-s}\int_0^{1-s} \left\{\min(x,y) - xy\right\}\,dF^-(x)dF^-(y).$$

Before proving the above theorem, we need the following lemma.

**Lemma 2.4.** *Under conditions of Theorem 2.24, we have*

$$\frac{\sqrt{k/n}\bar{F}^{-}(k/n)}{\sigma(k/n)} \to \sqrt{\frac{2-\alpha}{2}} I(\alpha < 2) \quad as \quad n \to \infty.$$

*Proof.* Write

$$\begin{aligned}
&\int_0^{1-k/n} \int_0^{1-k/n} \{\min(s,t) - st)\}\, dF^{-}(s) dF^{-}(t) \\
&= \int_0^{\bar{F}^{-}(k/n)} \int_0^{\bar{F}^{-}(k/n)} \left\{\min\left(F(s), F(t)\right) - F(s)F(t)\right\} dsdt \\
&= 2\int_0^{\bar{F}^{-}(k/n)} \int_0^{s} F(t)\bar{F}(s)\, dtds \\
&= 2\int_0^{\bar{F}^{-}(k/n)} s\bar{F}(s)\, ds - 2\int_0^{\bar{F}^{-}(k/n)} \int_0^{s} \bar{F}(t)\bar{F}(s)\, dtds \\
&= I_1 - I_2.
\end{aligned}$$

Then, for $1 < \alpha < 2$, we have as $\to \infty$

$$\frac{I_1}{\frac{k}{n}(\bar{F}^{-}(k/n))^2} = 2\int_0^1 s\frac{\bar{F}(\bar{F}^{-}(\frac{k}{n})s)}{\bar{F}(\bar{F}^{-}(\frac{k}{n}))}\, ds \to 2\int_0^1 s^{1-\alpha}\, ds = \frac{2}{2-\alpha} \tag{2.86}$$

and

$$\frac{I_2}{\frac{k}{n}(\bar{F}^{-}(k/n))^2} \le \frac{2\int_0^{\infty} \mu\bar{F}(s)\, ds}{\frac{k}{n}(\bar{F}^{-}(k/n))^2} = \frac{2\mu^2}{\frac{k}{n}(\bar{F}^{-}(k/n))^2} \to 0. \tag{2.87}$$

When $\alpha = 2$, by noting that $\rho > 0$ implies $\bar{F}(x) \sim cx^{-\alpha}$ for some $c > 0$ as $x \to \infty$, we have

$$\lim_{n\to\infty} \sqrt{k/n}\bar{F}^{-}(k/n) \text{ is finite and } \lim_{n\to\infty} \sigma(k/n) = \infty,$$

which imply that

$$\frac{\sqrt{k/n}\bar{F}^{-}(k/n)}{\sigma(k/n)} \to 0 \quad \text{as} \quad n \to \infty. \tag{2.88}$$

When $\alpha > 2$, we have

$$\lim_{n\to\infty} \sqrt{k/n}\bar{F}^{-}(k/n) = 0 \text{ and } \lim_{n\to\infty} \sigma(k/n) = \sigma(0) \in (0, \infty),$$

which imply that (2.88) still holds. Hence the lemma follows from (2.86)–(2.88). □

*Proof of Theorem 2.24.* Put $U_i = F(X_i)$ for $i = 1, \cdots, n$ and write

$$\begin{aligned}\hat{\mu}_1 - \mu_1 &= \int_0^{U_{n,1}} \{x - G_n(x)\}\, dF^-(x) + \int_{U_{n,1}}^{U_{n,n-k}} \{x - G_n(x)\}\, dF^-(x) \\ &\quad + \int_{1-k/n}^{U_{n,n-k}} (1-x)\, dF^-(x) \\ &= I_1 + I_2 + I_3.\end{aligned}$$

Then $I_1 = O_p(1/n)$ and

$$\begin{aligned}I_3 &= \frac{k}{n}\left\{F^-(U_{n,n-k}) - F^-(1-k/n)\right\}\left\{1 + o_p(1)\right\} \\ &= \frac{k}{n}\bar{F}^-(\frac{k}{n})\left\{(\frac{n}{k}(1-U_{n,n-k}))^{-1/\alpha} - 1\right\}\left\{1 + o_p(1)\right\} \\ &= -\frac{1}{\alpha}\frac{\sqrt{k}}{n}\bar{F}^-(\frac{k}{n})\sqrt{\frac{n}{k}}B_n(1-k/n)\left\{1 + o_p(1)\right\}.\end{aligned}$$

It follows from (2.26) that for any $\delta \in (0, 1/4)$

$$\begin{aligned}&|\sqrt{n}I_2 + \int_{U_{n,1}}^{U_{n,n-k}} B_n(x)\, dF^-(x)| \\ &= \left\{n^{-\delta}\int_{1/n}^{1-k/n} x^{1/2-\delta}(1-x)^{1/2-\delta}\, dF^-(x)\right\} O_p(1) \\ &= O_p\big(n^{-\delta}\int_{1/2}^{1-k/n}(1-x)^{1/2-\delta}\, dF^-(x)\big) \\ &= O_p\big(n^{-\delta}(\frac{k}{n})^{1/2-1/\alpha-\delta}\big) \\ &= o_p\big(\sqrt{\frac{k}{n}}\bar{F}^-(\frac{k}{n})\big).\end{aligned}$$

Using the above equations and Lemma 2.4, we have

$$\begin{aligned}&\frac{\sqrt{n}}{\sigma(k/n)}\left\{\hat{\mu}_1 - \mu_1\right\} \\ &= -\frac{\int_0^{1-k/n} B_n(x)\, dF^-(x)}{\sigma(k/n)} - \frac{1}{\alpha}\sqrt{\frac{2-\alpha}{2}}I(\alpha < 2)\sqrt{\frac{n}{k}}B_n(1-\frac{k}{n}) + o_p(1).\end{aligned} \tag{2.89}$$

Note that

$$\begin{aligned}
\mu_2 &= \frac{k}{n}\bar{F}^-(k/n)\left\{\int_1^\infty x^{-\alpha}\,dx + O(A(k/n))\right\}\\
&= \frac{k}{n}\frac{\bar{F}^-(k/n)}{\alpha-1} + \frac{k}{n}\bar{F}^-(k/n)O(A(k/n)),\\
\hat{\mu}_2 - \frac{k}{n}\frac{\bar{F}^-(k/n)}{\alpha-1} &= \frac{k}{n}\frac{X_{n,n-k}-\bar{F}^-(k/n)}{\hat{\alpha}(k)-1} - \frac{k}{n}\bar{F}^-(k/n)\left\{\frac{1}{\alpha-1}-\frac{1}{\hat{\alpha}(k)-1}\right\},\\
\sqrt{n}\frac{\frac{k}{n}\{X_{n,n-k}-\bar{F}^-(k/n)\}}{\sqrt{k/n}\bar{F}^-(k/n)} &= \sqrt{k}\left\{(\frac{n}{k}(1-U_{n,n-k}))^{-1/\alpha} - 1 + O_p(A(k/n))\right\}\\
&= -\frac{1}{\alpha}\sqrt{k}\left\{\frac{n}{k}(1-U_{n,n-k})-1\right\} + o_p(1)\\
&= -\frac{1}{\alpha}\sqrt{n/k}B_n(1-\frac{k}{n}) + o_p(1),\\
\sqrt{n}\frac{\frac{k}{n}\bar{F}^-(k/n)\{\hat{\alpha}(k)-\alpha\}}{\sqrt{k/n}\bar{F}^-(k/n)} &= \alpha\sqrt{n/k}\int_0^1\left\{\frac{B_n(1-\frac{k}{n}s)}{s} - B_n(1-\frac{k}{n})\right\}ds\\
&\quad + o_p(1),
\end{aligned}$$

which imply that

$$\begin{aligned}
\frac{\sqrt{n}\{\hat{\mu}_2-\mu_2\}}{\sigma(k/n)} &= -\frac{1}{\alpha(\alpha-1)}\sqrt{\frac{2-\alpha}{2}}I(\alpha<2)\sqrt{\frac{n}{k}}B_n(1-\frac{k}{n})\\
-\frac{\alpha}{(\alpha-1)^2}\sqrt{\frac{2-\alpha}{2}}I(\alpha<2)&\sqrt{\frac{n}{k}}\int_0^1\left\{\frac{B_n(1-\frac{k}{n}s)}{s} - B_n(1-\frac{k}{n})\right\}ds + o_p(1).
\end{aligned} \tag{2.90}$$

Hence it follows from (2.89) and (2.90) that

$$\begin{aligned}
&\frac{\sqrt{n}}{\sigma(k/n)}\{\hat{\mu}_1+\hat{\mu}_2-\mu\}\\
&= -\frac{\int_0^{1-k/n}B_n(x)\,dF^-(x)}{\sigma(k/n)} - \frac{1}{\alpha-1}\sqrt{\frac{2-\alpha}{2}}I(\alpha<2)\sqrt{\frac{n}{k}}B_n(1-\frac{k}{n})\\
&\quad -\frac{\alpha}{(\alpha-1)^2}\sqrt{\frac{2-\alpha}{2}}I(\alpha<2)\sqrt{\frac{n}{k}}\int_0^1\left\{\frac{B_n(1-\frac{k}{n}s)}{s} - B_n(1-\frac{k}{n})\right\}ds + o_p(1),
\end{aligned}$$

which implies the theorem by noting that

$$E\left\{\frac{\int_0^{1-k/n} B_n(s)\,dF^-(s)}{\sigma(k/n)}\right\}^2 = 1,\ E\left\{\sqrt{\frac{n}{k}}B_n(1-\frac{n}{k})\right\}^2 = 1-\frac{k}{n} \to 1,$$

$$E\left\{\sqrt{\frac{n}{k}}\int_0^1 \Big(\frac{B_n(1-\frac{k}{n}s)}{s} - B_n(1-\frac{k}{n})\Big)\,ds\right\}^2 \to 1,$$

$$E\left\{\sqrt{\frac{n}{k}}B_n(1-\frac{k}{n})\sqrt{\frac{n}{k}}\int_0^1 \Big(\frac{B_n(1-\frac{k}{n}s)}{s} - B_n(1-\frac{k}{n})\Big)\,ds\right\} = 0,$$

$$E\left\{\sqrt{\frac{n}{k}}B_n(1-\frac{k}{n})\frac{\int_0^{1-k/n} B_n(x)\,dF^-(x)}{\sigma(k/n)}\right\} = \frac{\sqrt{k/n}}{\sigma(k/n)}\int_0^{1-k/n} x\,dF^-(x)$$

$$= \frac{\sqrt{k/n}}{\sigma(k/n)}\left\{(1-\frac{k}{n})\bar{F}^-(\frac{k}{n}) - \int_0^{1-k/n} F^-(x)\,dx\right\} = \frac{\sqrt{k/n}}{\sigma(k/n)}\left\{\bar{F}^-(\frac{k}{n}) + O(1)\right\}$$

$$\to \sqrt{\frac{2-\alpha}{2}}I(\alpha < 2),$$

and

$$E\left\{\frac{\int_0^{1-k/n} B_n(s)\,dF^-(s)}{\sigma(k/n)}\sqrt{\frac{n}{k}}\int_0^1 \Big(\frac{B_n(1-\frac{k}{n}s)}{s} - B_n(1-\frac{k}{n})\Big)\,ds\right\} = 0. \qquad \square$$

Theorem 2.24 can be employed to construct a confidence interval for $E(X)$ via estimating $\alpha$ and $\sigma$ without knowing whether the random variable $X_i$ has a finite or an infinite variance. Here we introduce an empirical likelihood method without estimating the asymptotic variance.

Put $\delta_i = I(X_i \geq X_{n,n-k})$, $q = F^-(k/n)$, $c = \frac{k}{n}X^{\alpha}_{n,n-k}$, and write the log-likelihood function as

$$l_0(\alpha, p_1, \cdots, p_n) = \sum_{\delta_i=1}\left\{\log(\frac{k}{n}) + \alpha\log X_{n,n-k} + \log\alpha - (\alpha+1)\log X_i\right\} + \sum_{\delta_i=0}\log(p_i).$$

This is the sum of the logarithm of parametric likelihood for data in the upper tail and the logarithm of non-parametric likelihood for data below $X_{n,n-k}$. By considering the log-likelihood function $l_0$ subject to

$$\alpha > 1, \quad p_1 \geq 0, \cdots, p_n \geq 0, \tag{2.91}$$

$$\sum_{\delta_i=0} p_i = 1 - \frac{k}{n}, \tag{2.92}$$

and

$$\sum_{\delta_i=0} p_i X_i = \mu - \frac{\alpha}{\alpha-1}\frac{k}{n}q, \tag{2.93}$$

we follow the semiparametric likelihood ratio method in Qin and Wong [90] to obtain an interval for $\mu$.

First we maximize $l_0$ under constraints (2.91) and (2.92). It follows from the Lagrange multiplier technique that

$$\bar{\alpha} = \left\{ \frac{1}{k}\sum_{i=1}^{k} \log \frac{X_{n,n-i+1}}{X_{n,n-k}} \right\}^{-1}, \quad \bar{p}_i = \frac{1}{n} \text{ for } \delta_i = 0.$$

Next we maximize $l_0$ subject to (2.91)–(2.93). Define

$$w_i(\mu,\alpha) = X_i - \frac{\mu - \frac{\alpha}{\alpha-1}\frac{k}{n}q}{1-k/n}, \quad \lambda(\alpha) = (\frac{1}{\bar{\alpha}} - \frac{1}{\alpha})\frac{(\alpha-1)^2}{q},$$

and let $\hat{\alpha} = \hat{\alpha}(\mu)$ denote the solution to

$$\sum_{\delta_i=1} \frac{w_i(\mu,\alpha)}{1+\lambda(\alpha)w_i(\mu,\alpha)} = 0. \tag{2.94}$$

Therefore the values maximizing $l_0$ subject to (2.91)–(2.93) are

$$\hat{\lambda} = \hat{\lambda}(\hat{\alpha}), \ \hat{p}_i = \left\{ n\big(1+\hat{\lambda}w_i(\mu,\hat{\alpha})\big)\right\}^{-1} \text{ for } \delta_i = 0.$$

Hence the semiparametric likelihood ratio multiplied by $-2$ is

$$l(\mu) = -2\left\{ l_0(\hat{\alpha},\hat{p}_1,\cdots,\hat{p}_n) - l_0(\bar{\alpha},\bar{p}_1,\cdots,\bar{p}_n)\right\}.$$

**Theorem 2.25.** *Under conditions of (2.4) with $\rho > 0$, (2.28),*

$$\sqrt{k}\frac{A(k/n)}{(k/n)^{1/\alpha}\bar{F}^{-}(k/n)} = O(n^{-\delta_0}) \textit{ for some } \delta_0 > 0,$$

*$l(\mu_0)$ converges in distribution to a chi-squared distribution with one degree of freedom as $n \to \infty$, where $\mu_0$ denotes the true value of $\mu$.*

*Proof.* Let $\eta = (\alpha, p_1, \cdots, p_n)^T$. The key idea in our proof is to expand $l_0(\hat{\eta})$ around $\bar{\eta}$ and then to derive the limit for $\hat{\alpha} - \bar{\alpha}$.

Using the same notation and similar expansions in proving Theorem 2.24, Lemma 2.4 and Taylor expansions, we can show that

$$\hat{\alpha}(\mu_0) - \alpha_0 = O_p(k^{-1/2}),$$
$$\sqrt{k}(\bar{\alpha} - \alpha_0) = \alpha_0 \int_0^1 \left\{ \frac{B_n(1 - \frac{k}{n}s)}{s} - B_n(1 - \frac{k}{n}) \right\} ds + o_p(1),$$

$$\begin{aligned}
&\frac{\sqrt{n}}{\sigma(k/n)} \frac{1}{n} \sum_{\delta_i=1} w_i(\mu_0, \hat{\alpha}(\mu_0)) \\
= &\frac{\sqrt{n}}{\sigma(k/n)} \Big\{ \int_0^{X_{n,n-k}} x\, d\big(F_n(x) - F(x)\big) - \int_{X_{n,n-k}}^{\bar{F}^-(k/n)} x\, dF(x) \\
&- \int_{\bar{F}^-(k/n)}^{\infty} x\, dF(x) + \frac{\hat{\alpha}(\mu_0)}{\hat{\alpha}(\mu_0) - 1} \frac{k}{n} q \Big\} \\
= &\frac{\sqrt{n}}{\sigma(k/n)} \Big\{ X_{n,n-k}\big(F_n(X_{n,n-k}) - F(X_{n,n-k})\big) - \int_0^{X_{n,n-k}} \big(F_n(x) - F(x)\big)\, dx \\
&- \int_{X_{n,n-k}}^{\bar{F}^-(k/n)} x\, dF(x) + \frac{k}{n} q \int_1^{\infty} x\, d\big(\frac{\bar{F}(\bar{F}^-(k/n)x)}{\bar{F}(\bar{F}^-(k/n))} - x^{-\alpha_0}\big) \\
&+ \big(\frac{\hat{\alpha}(\mu_0)}{\hat{\alpha}(\mu_0) - 1} - \frac{\alpha_0}{\alpha_0 - 1}\big) \frac{k}{n} q \Big\} \\
= &\frac{\sqrt{n}}{\sigma(k/n)} \Big\{ - \int_0^{X_{n,n-k}} \big(F_n(x) - F(x)\big)\, dx + \big(\frac{\hat{\alpha}(\mu_0)}{\hat{\alpha}(\mu_0) - 1} - \frac{\alpha_0}{\alpha_0 - 1}\big) \frac{k}{n} q \Big\} \\
&+ o_p(1) \\
= &- \frac{\int_0^{1-k/n} B_n(s)\, dF^-(s)}{\sigma(k/n)} - \frac{\sqrt{k}(\bar{\alpha} - \alpha_0)}{(\alpha_0 - 1)^2} \delta - \frac{\sqrt{k}(\hat{\alpha}(\mu_0) - \bar{\alpha})}{(\alpha_0 - 1)^2} \delta + o_p(1),
\end{aligned}$$

$$\begin{cases} \sqrt{k}\hat{\lambda}(\hat{\alpha}(\mu_0)) q = \sqrt{k}(\hat{\alpha}(\mu_0) - \bar{\alpha}) \dfrac{(\alpha_0 - 1)^2}{\alpha_0^2} + o_p(1), \\ \hat{\lambda}(\hat{\alpha}(\mu_0)) = \dfrac{n}{\sigma^2(k/n)} \dfrac{1}{n} \displaystyle\sum_{\delta_i=1} w_i(\mu_0, \hat{\alpha}(\mu_0)), \end{cases} \tag{2.95}$$

where $\delta = \sqrt{\frac{2-\alpha_0}{2}} I(\alpha_0 < 2)$, $\alpha_0$ denotes the true value of $\alpha$ and $\sigma(s)$ is defined in Theorem 2.24. Put

$$A_n = -\frac{\int_0^{1-k/n} B_n(s)\, dF^-(s)}{\sigma(k/n)} - \delta \frac{\alpha_0}{(\alpha_0-1)^2} \int_0^1 \left\{ \frac{B_n(1-\frac{k}{n}s)}{s} - B_n(1-\frac{k}{n}) \right\} ds.$$

Therefore it follows from the above two expressions for $\hat{\lambda}(\hat{\alpha}(\mu_0))$ in (2.95) that

$$\sqrt{k}\big\{\hat{\alpha}(\mu_0) - \bar{\alpha}\big\} \frac{(\alpha_0-1)^2}{\alpha_0^2} = \delta A_n - \delta^2 \frac{\sqrt{k}\{\hat{\alpha}(\mu_0) - \bar{\alpha}\}}{(\alpha_0-1)^2} + o_p(1),$$

i.e.,

$$\sqrt{k}\big\{\hat{\alpha}(\mu_0) - \bar{\alpha}\big\} = \frac{(\alpha_0-1)^2 \delta A_n}{\delta^2 + (\alpha_0-1)^4 \alpha_0^{-2}} + o_p(1).$$

Further we have

$$\frac{\sqrt{n}}{\sigma(k/n)} \frac{1}{n} \sum_{\delta_i=1} w_i(\mu_0, \hat{\alpha}(\mu_0)) = \frac{A_n}{1+\delta^2 \alpha_0^2 (\alpha_0-1)^{-4}} + o_p(1).$$

Expanding $l_0(\hat{\eta})$ around $\bar{\eta}$, we have

$$\begin{aligned} l(\mu_0) &= -2 \left\{ \frac{1}{2} \frac{\partial^2 l_0(\bar{\eta})}{\partial \alpha^2} (\hat{\alpha}(\mu_0) - \bar{\alpha})^2 - \frac{n}{2\sigma^2(k/n)} \Big( \frac{1}{n} \sum_{\delta_i=0} w_i(\mu_0, \hat{\alpha}(\mu_0)) \Big)^2 \right\} \\ &\quad + o_p(1) \\ &= \frac{A_n^2}{1+\delta^2 \alpha_0^2 (\alpha_0-1)^{-4}} + o_p(1), \end{aligned}$$

which implies the theorem by noting that $A_n$ has a normal distribution with mean zero and variance $1+\delta^2\alpha_0^2(\alpha_0-1)^{-4}$. □

## 2.8 EXPECTED SHORTFALL

Suppose $X, X_1, \cdots, X_n$ are independent and identically distributed random variables with a continuous distribution function $F$ satisfying (2.1). In addition to Value-at-Risk (i.e., quantile), another commonly used risk measure in insurance and banking industry is the so-called expected shortfall at level $1-q \in (0,1)$, which is defined as

$$\rho(q) = E\big(X | F(X) > 1-q\big).$$

An obvious nonparametric estimator is

$$\rho_n(q) = \frac{1}{nq} \sum_{i=1}^{n} X_i I(X_i \geq X_{n,n-[nq]}),$$

where $X_{n,1} \leq \cdots \leq X_{n,n}$ are the order statistics of $X_1, \cdots, X_n$. Like the study of quantiles, we consider the cases of fixed level and extreme level.

**Case 1: Fixed level.** Here $q \in (0,1)$ is fixed. It is known that the asymptotic distribution of $\rho_n(q)$ will be nonnormal if $\alpha < 2$. Like the study of estimating a mean, we could estimate the expected shortfall by combining a parametric estimator for the tail part and a nonparametric estimator for the middle part as follows.

Write

$$\begin{aligned}
\rho(q) &= q^{-1} \int_{\bar{F}^{-}(q)}^{\bar{F}^{-}(k/n)} x\, dF(x) + q^{-1} \int_{\bar{F}^{-}(k/n)}^{\infty} x\, dF(x) \\
&= \bar{F}^{-}(q) + q^{-1} \int_{\bar{F}^{-}(q)}^{\bar{F}^{-}(k/n)} \bar{F}(x)\, dx + q^{-1} \int_{\bar{F}^{-}(k/n)}^{\infty} \bar{F}(x)\, dx \\
&=: \rho_1 + \rho_2 + \rho_3,
\end{aligned}$$

which motivates us to estimate $\rho_1, \rho_2, \rho_3$ by

$$\hat{\rho}_1 = X_{n,n-[nq]}, \quad \hat{\rho}_2 = q^{-1} \int_{X_{n,n-[nq]}}^{X_{n,n-k}} \bar{F}_n(x)\, dx, \quad \hat{\rho}_3 = \frac{k}{nq} \frac{X_{n,n-k}}{\hat{\alpha}(k) - 1},$$

respectively, and accordingly to estimate $\rho(q)$ by

$$\hat{\rho}(q) = \hat{\rho}_1 + \hat{\rho}_2 + \hat{\rho}_3,$$

where $F_n(x) = \frac{1}{n} \sum_{i=1}^{n} I(X_i \leq x)$ and $\hat{\alpha}(k)$ is the Hill estimator in (2.31).

**Theorem 2.26.** *Under conditions of (2.4) with $\rho > 0$, (2.28) and*

$$\lim_{n\to\infty} \sqrt{k} \frac{A(k/n)}{(k/n)^{1/\alpha} \bar{F}^{-}(k/n)} = 0,$$

*we have for a fixed $q \in (0,1)$*

$$\frac{\sqrt{n}}{\sigma(k/n)} \left\{ \hat{\rho}(q) - \rho(q) \right\} \xrightarrow{d} N(0, \sigma_0^2) \quad \textit{as} \quad n \to \infty,$$

*where* $\sigma^2(s) = \int_{1-q}^{1-s}\int_{1-q}^{1-s}\{\min(x,y) - xy\}\, dF^-(x)dF^-(y)$ *and*

$$\sigma_0^2 = 1 + \left\{\frac{(1-q)^3 q}{f^2(\bar{F}^-(q))\sigma^2(0)} - \frac{2(1-q)^2}{f(\bar{F}^-(q))\sigma^2(0)}\int_0^q s\, d\bar{F}^-(s)\right\} I(\alpha > 2)$$
$$+ \left\{\frac{2-\alpha}{2\alpha^2} + \frac{2-\alpha}{2q^2\alpha^2(\alpha-1)^2} + \frac{\alpha^2(2-\alpha)}{2q^2(\alpha-1)^4} + \frac{2-\alpha}{\alpha} + \frac{2-\alpha}{q\alpha^2(\alpha-1)}\right.$$
$$\left. + \frac{2-\alpha}{q\alpha(\alpha-1)}\right\} I(\alpha < 2).$$

**Remark 2.10.** When $q = 1$, the above $\sigma_0^2$ becomes the same as the asymptotic variance in Theorem 2.24.

Before proving the above theorem, we need a lemma.

**Lemma 2.5.** *Under conditions of Theorem 2.26, we have*

$$\frac{\sqrt{k/n}\bar{F}^-(k/n)}{\sigma(k/n)} \to \sqrt{\frac{2-\alpha}{2}} I(\alpha < 2) \quad as \quad n \to \infty.$$

*Proof.* Write

$$\int_{1-q}^{1-k/n}\int_{1-q}^{1-k/n}\{\min(s,t) - st)\}\, dF^-(s)dF^-(t)$$
$$= 2\int_{\bar{F}^-(q)}^{\bar{F}^-(k/n)}\int_{\bar{F}^-(q)}^{s} F(t)\bar{F}(s)\, dtds$$
$$= 2\int_{\bar{F}^-(q)}^{\bar{F}^-(k/n)} (s - \bar{F}^-(q))\bar{F}(s)\, ds - 2\int_{\bar{F}^-(q)}^{\bar{F}^-(k/n)}\int_{\bar{F}^-(q)}^{s} \bar{F}(t)\bar{F}(s)\, dtds$$
$$= I_1 - I_2.$$

Then, for $1 < \alpha < 2$ we have

$$\frac{I_1}{\frac{k}{n}(\bar{F}^-(k/n))^2} = 2\int_{\bar{F}^-(q)/\bar{F}^-(k/n)}^{1} s\frac{\bar{F}(\bar{F}^-(\frac{k}{n})s)}{\bar{F}(\bar{F}^-(\frac{k}{n}))}\, ds \to 2\int_0^1 s^{1-\alpha}\, ds = \frac{2}{2-\alpha} \tag{2.96}$$

and

$$\frac{I_2}{\frac{k}{n}(\bar{F}^-(k/n))^2} \le \frac{2\int_{\bar{F}^-(q)}^{\infty}\bar{\mu}\bar{F}(s)\, ds}{\frac{k}{n}(\bar{F}^-(k/n))^2} = \frac{2\bar{\mu}^2}{\frac{k}{n}(\bar{F}^-(k/n))^2} \to 0, \tag{2.97}$$

where $\bar{\mu} = \int_{\bar{F}^-(q)}^{\infty}\bar{F}(s)\, ds < \infty$.

When $\alpha = 2$, by noting that $\rho > 0$ implies $\bar{F}(x) \sim cx^{-\alpha}$ for some $c > 0$ as $x \to \infty$, we have

$$\lim_{n\to\infty} \sqrt{k/n}\bar{F}^{-}(k/n) \text{ is finite and } \lim_{n\to\infty} \sigma(k/n) = \infty,$$

which imply that

$$\frac{\sqrt{k/n}\bar{F}^{-}(k/n)}{\sigma(k/n)} \to 0 \quad \text{as} \quad n \to \infty. \tag{2.98}$$

When $\alpha > 2$, we have

$$\lim_{n\to\infty} \sqrt{k/n}\bar{F}^{-}(k/n) = 0 \text{ and } \lim_{n\to\infty} \sigma(k/n) = \sigma(0) \in (0, \infty),$$

which imply that (2.98) still holds. Hence the lemma follows from (2.96)–(2.98). □

*Proof of Theorem 2.26.* Put $U_i = F(X_i)$ for $i = 1, \cdots, n$ and write

$$\begin{aligned}
\hat{\rho}_2 - \rho_2 &= \int_{U_{n,n-[nq]}}^{U_{n,n-k}} \{1 - G_n(x)\}\, dF^{-}(x) - \int_{1-q}^{1-k/n} \{1 - x\}\, dF^{-}(x) \\
&= \int_{U_{n,n-[nq]}}^{1-q} \{1 - G_n(x)\}\, dF^{-}(x) + \int_{1-k/n}^{U_{n,n-k}} \{1 - G_n(x)\}\, dF^{-}(x) \\
&\quad + \textstyle\int_{1-q}^{1-k/n} (x - G_n(x))\, dF^{-}(x) \\
&= I_1 + I_2 + I_3.
\end{aligned}$$

Then by (2.27), (2.30) and (2.9), we have

$$\begin{aligned}
I_1 &= q\left\{F^{-}(1-q) - F^{-}(U_{n,n-[nq]})\right\}\left\{1 + o_p(1)\right\} \\
&= \frac{q}{f(\bar{F}^{-}(q))}\left\{1 - q - U_{n,n-[nq]}\right\}\left\{1 + o_p(1)\right\} \\
&= \frac{q}{f(\bar{F}^{-}(q))}\frac{B_n(1-q)}{\sqrt{n}}\left\{1 + o_p(1)\right\},
\end{aligned}$$

and

$$
\begin{aligned}
I_2 &= \frac{k}{n}\left\{F^{-}(U_{n,n-k}) - F^{-}(1-k/n)\right\}\left\{1+o_p(1)\right\} \\
&= \frac{k}{n}\bar{F}^{-}(\frac{k}{n})\left\{(\frac{n}{k}(1-U_{n,n-k}))^{-1/\alpha} - 1\right\}\left\{1+o_p(1)\right\} \\
&= -\frac{1}{\alpha}\frac{\sqrt{k}}{n}\bar{F}^{-}(\frac{k}{n})\sqrt{\frac{n}{k}}B_n(1-k/n)\left\{1+o_p(1)\right\}.
\end{aligned}
$$

It follows from (2.29) that for any $\delta \in (0, 1/4)$

$$
\begin{aligned}
&|\sqrt{n}I_3 + \int_{1-q}^{1-k/n} B_n(x)\, dF^{-}(x)| \\
&= \left\{n^{-\delta}\int_{1-q}^{1-k/n} x^{1/2-\delta}(1-x)^{1/2-\delta}\, dF^{-}(x)\right\} O_p(1) \\
&= O_p\big(n^{-\delta}\int_{1-q}^{1-k/n}(1-x)^{1/2-\delta}\, dF^{-}(x)\big) = O_p\big(n^{-\delta}(\frac{k}{n})^{1/2-1/\alpha-\delta}\big) \\
&= o_p\big(\sqrt{k/n}\bar{F}^{-}(k/n)\big).
\end{aligned}
$$

Using the above equations, Lemma 2.5 and the fact that $\lim_{n\to\infty}\sigma(k/n) = \infty$ when $\alpha \le 2$, we have

$$
\begin{aligned}
&\frac{\sqrt{n}}{\sigma(k/n)}\left\{\hat{\rho}_2 - \rho_2\right\} \\
&= \frac{q}{f(\bar{F}^{-}(q))\sigma(0)}B_n(1-q)I(\alpha > 2) - \frac{1}{\alpha}\sqrt{\frac{2-\alpha}{2}}\sqrt{\frac{n}{k}}B_n(1-\frac{k}{n})I(\alpha < 2) \\
&\quad - \frac{\int_{1-q}^{1-k/n} B_n(x)\, dF^{-}(x)}{\sigma(k/n)} + o_p(1).
\end{aligned}
\tag{2.99}
$$

Note that

$$
\begin{aligned}
\frac{\sqrt{n}}{\sigma(k/n)}\left\{\hat{\rho}_1 - \rho_1\right\} &= \frac{\sqrt{n}}{\sigma(k/n)}\left\{\bar{F}^{-}(1-U_{n,n-[nq]}) - \bar{F}^{-}(q)\right\} \\
&= -\frac{\sqrt{n}}{\sigma(k/n)}\frac{1}{f(\bar{F}^{-}(q))}\left\{1-U_{n,n-[nq]} - q\right\}\left\{1+o_p(1)\right\} \\
&= -\frac{\sqrt{n}}{\sigma(k/n)}\frac{1}{f(\bar{F}^{-}(q))}\frac{B_n(1-q)}{\sqrt{n}}\left\{1+o_p(1)\right\} \\
&= -\frac{1}{f(\bar{F}^{-}(q))\sigma(0)}B_n(1-q)I(\alpha > 2)(1+o_p(1)),
\end{aligned}
\tag{2.100}
$$

$$\begin{aligned}\rho_3 &= q^{-1}\frac{k}{n}\bar{F}^{-}(k/n)\left\{\int_1^\infty x^{-\alpha}\,dx + O\big(|A(k/n)|\big)\right\}\\ &= q^{-1}\frac{k}{n}\bar{F}^{-}(k/n)\frac{1}{\alpha-1} + O\big(\frac{k}{n}\bar{F}^{-}(k/n)|A(k/n)|\big),\end{aligned}$$

$$\begin{aligned}\hat{\rho}_3 - q^{-1}\tfrac{k}{n}\tfrac{\bar{F}^{-}(k/n)}{\alpha-1} = &\; q^{-1}\tfrac{k}{n}\tfrac{X_{n,n-k}-\bar{F}^{-}(k/n)}{\hat{\alpha}(k)-1}\\ &- q^{-1}\frac{k}{n}\bar{F}^{-}(k/n)\left\{\frac{1}{\alpha-1}-\frac{1}{\hat{\alpha}(k)-1}\right\},\end{aligned}$$

$$\begin{aligned}\sqrt{n}\frac{\frac{k}{n}\{X_{n,n-k}-\bar{F}^{-}(k/n)\}}{\sqrt{k/n}\bar{F}^{-}(k/n)} &= \sqrt{k}\left\{(\frac{n}{k}(1-U_{n,n-k}))^{-1/\alpha} - 1 + O_p\big(|A(k/n)|\big)\right\}\\ &= -\frac{1}{\alpha}\sqrt{k}\left\{\frac{n}{k}(1-U_{n,n-k})-1\right\}+o_p(1)\\ &= -\frac{1}{\alpha}\sqrt{\frac{n}{k}}B_n(1-\frac{k}{n})+o_p(1),\end{aligned}$$

and

$$\sqrt{n}\frac{\frac{k}{n}\bar{F}^{-}(k/n)\{\hat{\alpha}(k)-\alpha\}}{\sqrt{k/n}\bar{F}^{-}(k/n)} = \alpha\sqrt{\frac{n}{k}}\int_0^1\left\{\frac{B_n(1-\frac{k}{n}s)}{s} - B_n(1-\frac{k}{n})\right\}ds + o_p(1),$$

which imply that

$$\begin{aligned}&\frac{\sqrt{n}\{\hat{\rho}_3-\rho_3\}}{\sigma(k/n)}\\ &= -\frac{1}{q\alpha(\alpha-1)}\sqrt{\frac{2-\alpha}{2}}I(\alpha<2)\sqrt{\frac{n}{k}}B_n(1-\frac{k}{n})\\ &\quad - \frac{\alpha}{q(\alpha-1)^2}\sqrt{\frac{2-\alpha}{2}}I(\alpha<2)\sqrt{\frac{n}{k}}\int_0^1\left\{\frac{B_n(1-\frac{k}{n}s)}{s}-B_n(1-\frac{k}{n})\right\}ds\\ &\quad + o_p(1).\end{aligned} \tag{2.101}$$

Hence it follows from (2.99), (2.100) and (2.101) that

$$\begin{aligned}&\frac{\sqrt{n}}{\sigma(k/n)}\{\hat{\rho}(q)-\rho(q)\}\\ &= -\frac{1-q}{f(\bar{F}^{-}(q))\sigma(0)}B_n(1-q)I(\alpha>2) - \frac{1}{\alpha}\sqrt{\frac{2-\alpha}{2}}I(\alpha<2)\sqrt{\frac{n}{k}}B_n(1-\frac{k}{n})\end{aligned}$$

$$
\begin{aligned}
&-\frac{\int_{1-q}^{1-k/n} B_n(s)\,dF^-(s)}{\sigma(k/n)} - \frac{1}{q\alpha(\alpha-1)}\sqrt{\frac{2-\alpha}{2}}I(\alpha<2)\sqrt{\frac{n}{k}}B_n(1-\frac{k}{n})\\
&-\frac{\alpha}{q(\alpha-1)^2}\sqrt{\frac{2-\alpha}{2}}I(\alpha<2)\sqrt{\frac{n}{k}}\int_0^1\left\{\frac{B_n(1-\frac{k}{n}s)}{s}-B_n(1-\frac{k}{n})\right\}ds\\
&+o_p(1),
\end{aligned}
$$

which implies the theorem by noting that

$$
E\left\{B_n(1-q)\right\}^2 = q(1-q), \quad E\left\{\sqrt{\frac{n}{k}}B_n(1-\frac{k}{n})\right\}^2 \to 1,
$$

$$
E\left\{\frac{\int_{1-q}^{1-k/n} B_n(s)\,dF^-(s)}{\sigma(k/n)}\right\}^2 = 1,
$$

$$
E\left\{\sqrt{\frac{n}{k}}\int_0^1 (\frac{B_n(1-\frac{k}{n}s)}{s}-B_n(1-\frac{k}{n}))\,ds\right\}^2 \to 1,
$$

$$
\begin{aligned}
&E\left\{B_n(1-q)\frac{\int_{1-q}^{1-k/n} B_n(x)\,dF^-(x)}{\sigma(k/n)}\right\}I(\alpha>2)\\
&=\frac{\int_{1-q}^{1-k/n}(1-q)(1-x)\,dF^-(x)}{\sigma(k/n)}I(\alpha>2)\\
&=-I(\alpha>2)\frac{1-q}{\sigma_0}\int_0^q x\,d\bar{F}^-(x)+o(1),
\end{aligned}
$$

$$
\begin{aligned}
&E\left\{\sqrt{\frac{n}{k}}B_n(1-\frac{k}{n})\frac{\int_{1-q}^{1-k/n} B_n(x)\,dF^-(x)}{\sigma(k/n)}\right\}\\
&=\frac{\sqrt{k/n}}{\sigma(k/n)}\int_{1-q}^{1-k/n} x\,dF^-(x)\\
&=\frac{\sqrt{k/n}}{\sigma(k/n)}\left\{(1-\frac{k}{n})\bar{F}^-(\frac{k}{n})-(1-q)\bar{F}^-(q)-\int_{1-q}^{1-k/n}F^-(x)\,dx\right\}\\
&=\frac{\sqrt{k/n}}{\sigma(k/n)}\left\{\bar{F}^-(\frac{k}{n})+O(1)\right\}\\
&\to\sqrt{\frac{2-\alpha}{2}}I(\alpha<2),
\end{aligned}
$$

$$
E\left\{\sqrt{\frac{n}{k}}B_n(1-\frac{k}{n})\sqrt{\frac{n}{k}}\int_0^1(\frac{B_n(1-\frac{k}{n}s)}{s}-B_n(1-\frac{k}{n}))\,ds\right\}=0
$$

and

$$E\left\{\frac{\int_{1-q}^{1-k/n} B_n(s)\,dF^-(s)}{\sigma(k/n)}\sqrt{\frac{n}{k}}\int_0^1\left(\frac{B_n(1-\frac{k}{n}s)}{s}-B_n(1-\frac{k}{n})\right)ds\right\}=0. \qquad \square$$

**Case 2: Extreme level.** Here we consider the case of $q=q_n\to 0$ and $nq_n\to d_0\in[0,\infty)$ as $n\to\infty$. In this case we need to extrapolate data. Note that

$$\begin{aligned}\rho(q)&=q^{-1}\int_{\bar F^-(q)}^{\infty} x\,dF(x)=-q^{-1}\int_{\bar F^-(q)}^{\infty} x\,d\bar F(x)\\&=\bar F^-(q)+\bar F^-(q)\int_1^{\infty}\frac{\bar F(\bar F^-(q)x)}{\bar F(\bar F^-(q))}\,dx\\&\sim \bar F^-(q)+\bar F^-(q)\int_1^{\infty}x^{-\alpha}\,dx=\bar F^-(q)\frac{\alpha}{\alpha-1},\end{aligned}$$

which motivates us to estimate $\rho(q)$ by

$$\tilde\rho(q)=X_{n,n-k}(nq/k)^{-1/\hat\alpha(k)}\frac{\hat\alpha(k)}{\hat\alpha(k)-1}.$$

**Theorem 2.27.** *Under conditions of (2.4) with $\rho>0$, (2.28), $q=q_n\to 0$, $nq\to d_0\in[0,\infty)$,*

$$\frac{\sqrt{k}A(k/n)}{(k/n)^{1/\alpha}\bar F^-(k/n)}\to\lambda\in\mathbb{R} \text{ and } \frac{\sqrt{k}}{\log(nq/k)}\to\infty \quad as \quad n\to\infty,$$

*we have*

$$\frac{\sqrt{k}}{\log(nq/k)}\log\frac{\tilde\rho(q)}{\rho(q)}\xrightarrow{d} N\left(\frac{\lambda}{1+\rho},\frac{1}{\alpha^2}\right) \quad as \quad n\to\infty.$$

*Proof.* Since $nq/k\to 0$, like the proof of Theorem 2.16, we can show that

$$\log\frac{\tilde\rho(q)}{\rho(q)}=\log(nq/k)\left\{-\frac{1}{\hat\alpha(k)}+\frac{1}{\alpha}\right\}\{1+o_p(1)\}. \tag{2.102}$$

Hence the theorem follows from (2.102) and Theorem 2.4. $\square$

**Remark 2.11.** The asymptotic distribution in Theorem 2.27 is the same as that for the high quantile estimator in Theorem 2.16. Therefore, data-driven methods for choosing the sample fraction $k$ in the Hill estimator $\hat\alpha(k)$ can be employed for the expected shortfall estimator $\tilde\rho(q)$ at an extreme level.

## 2.9 HAEZENDONCK–GOOVAERTS (H–G) RISK MEASURE

Let $\psi : [0, \infty] \to [0, \infty]$ be a convex function satisfying $\psi(0) = 0$, $\psi(1) = 1$ and $\psi(\infty) = \infty$, i.e., $\psi$ is a so-called normalized Young function. For a number $q \in (0, 1)$ and each $\beta > 0$, let $\alpha = \alpha(\beta)$ be a solution to

$$E\left\{\psi(\frac{(X-\beta)_+}{\alpha})\right\} = 1 - q, \tag{2.103}$$

where $x_+ = \max(x, 0)$. Then, Haezendonck and Goovaerts [49] proposed the so-called H–G risk measure at level $q$ as

$$\theta = \inf_{\beta > 0} \{\beta + \alpha(\beta)\}. \tag{2.104}$$

When $\psi(x) = x$, we have $\alpha(\beta) = \frac{1}{1-q} E\{(X - \beta)_+\}$ and $\theta = \frac{1}{1-q} E\{(X - F^-(q))_+\}$. Hence, in this case, the H–G risk measure equals the expected shortfall. In order to employ this risk measure in practice, an efficient statistical inference is needed. Ahn and Shyamalkumar [1] first proposed a nonparametric estimation and derived its asymptotic limit, which may be nonnormal when there are no enough moments, which depends on both the loss variable and the involved Young function $\psi$. When the limit is normal, Peng et al. [86] developed an empirical likelihood method to effectively construct an interval when the H–G risk measure is defined at a fixed level. Further, Wang and Peng [103] showed that this empirical likelihood method is still valid for an intermediate level, which leads to a unified interval estimator of the H–G risk measure at either a fixed level or an intermediate level.

The following theorem shows the asymptotic behavior of the H–G risk measure when the level $q$ goes to one.

**Theorem 2.28.** *Let $\psi(t)$ be strictly convex and continuously differentiable on $[0, \infty)$ with $\psi'(0+) = 0$, $\psi \in RV^0_{\beta_1}$ and $\psi \in RV^\infty_{\beta_2}$ with $1 < \beta_1, \beta_2 < \infty$. If $\bar{F} \in RV^\infty_{-\alpha}$ with some $\alpha \in \left(0, \min(\beta_1^{-1}, \beta_2^{-1})\right)$, then*

$$\lim_{q \to 1} \frac{\theta}{F^-(q)} = (1 + \frac{1}{\alpha\lambda})^{-1} \left(\int_0^\infty (1 + \frac{z}{\alpha\lambda})^{-\alpha} \, d\psi(z)\right)^{1/\alpha},$$

*where $\lambda > 0$ is the unique solution to the equation $E\big(\psi'(\lambda Y)\big) = E\big(\psi'(\lambda Y)\lambda Y\big)$ with $Y$ having the distribution function*

$$P(Y \le y) = 1 - \left(1 + y/\alpha\right)^{-\alpha} \quad \text{for} \quad y \ge 0.$$

*Proof.* See Corollary 6.1 of Tang and Yang [100]. □

The above theorem can be employed to estimate the H–G risk measure at an extreme level. Put $p = 1 - q$ and estimate the index $\alpha$ and the high quantile $F^{-}(q) = \bar{F}^{-}(p)$ by $\hat{\alpha}(k)$ in (2.31) and $\hat{x}_p$ in (2.56), respectively. Let $\hat{\lambda}(k)$ denote the solution to

$$\int_0^\infty \psi'(y)(1 + \frac{y}{\lambda\hat{\alpha}(k)})^{-\hat{\alpha}(k)-1}\, dy = \int_0^\infty \psi'(y)y(1 + \frac{y}{\lambda\hat{\alpha}(k)})^{-\hat{\alpha}(k)-1}\, dy. \quad (2.105)$$

Therefore we can estimate $\theta$ by

$$\hat{\theta}(k) = \left\{1 + \frac{1}{\hat{\alpha}(k)\hat{\lambda}(k)}\right\}^{-1} \left\{\int_0^\infty (1 + \frac{z}{\hat{\alpha}(k)\hat{\lambda}(k)})^{-\hat{\alpha}(k)}\, d\psi(z)\right\}^{1/\hat{\alpha}(k)} \hat{x}_p.$$

**Theorem 2.29.** *Under conditions of Theorem 2.16 with $\rho > 0$ and Theorem 2.28, we have*

$$\frac{\sqrt{k}}{\log(np/k)}\left\{\frac{\hat{\theta}(k)}{\theta} - 1\right\} \xrightarrow{d} N(\frac{\lambda}{1+\rho}, \frac{1}{\alpha^2}) \quad \text{as} \quad n \to \infty.$$

*Proof.* Following the proofs in Mao and Hu [71] by using Theorem 2.28, we can show that

$$\frac{\theta}{F^{-}(q)} = (1 + \frac{1}{\alpha\lambda})^{-1}\Big(\int_0^\infty (1 + \frac{z}{\alpha\lambda})^{-\alpha}\, d\psi(z)\Big)^{1/\alpha} + O_p\big(A(1-q)\big). \quad (2.106)$$

By (2.105) and Taylor expansions, we can show that

$$\hat{\lambda}(k) - \lambda = O_p(1/\sqrt{k}). \quad (2.107)$$

Hence the theorem follows from Theorem 2.16, (2.106), (2.107) and

$$\lim_{n\to\infty} \frac{A(1-q)}{A(k/n)\log(np/k)} = 0 \quad \text{due to} \quad \rho > 0.$$

□

# CHAPTER 3

# Heavy Tailed Dependent Data

Although losses in insurance are arguably independent and heavy tailed distributed, data in finance often exhibit dependence across time and some stylized facts such as heavy tails, skewness and heteroscedasticity. This chapter introduces techniques for inferring heavy tailed dependent data with a focus on heavy tailed time series models. A general extreme value theory for dependent sequences is available in the excellent book of Leadbetter et al. [67].

## 3.1 TAIL EMPIRICAL PROCESS AND TAIL QUANTILE PROCESS

Let $\{U_i\}_{i=-\infty}^{\infty}$ be a strictly stationary $\beta$-mixing sequence of uniformly distributed random variables on $[0,1]$, i.e.,

$$\beta(m) := \sup_{l\ge 1} E\left\{ \sup_{A\in\mathcal{B}_{l+m+1}^{\infty}} |P(A|\mathcal{B}_1^l) - P(A)| \right\} \to 0 \quad \text{as} \quad m\to\infty, \tag{3.1}$$

where $\mathcal{B}_1^l$ and $\mathcal{B}_{l+m+1}^{\infty}$ denote the $\sigma$-fields generated by $\{U_i\}_{i=1}^l$ and $\{U_i\}_{i=m+l+1}^{\infty}$, respectively. On the other hand, $\{U_i\}_{i=-\infty}^{\infty}$ is called a strictly $\alpha$-mixing sequence if

$$\alpha(m) = \sup\{|P(A\cap B) - P(A)P(B)| : A\in\mathcal{B}_1^l, B\in\mathcal{B}_{l+m+1}^{\infty}, l\ge 1\} \to 0 \quad \text{as} \quad m\to\infty.$$

As before, $U_{n,1}\le\cdots\le U_{n,n}$ denote the order statistics of $U_1,\cdots,U_n$ and the tail empirical process and the tail quantile process are defined as

$$\alpha_{n,k}(s) = \sqrt{k}\left\{\frac{1}{k}\sum_{i=1}^n I(U_i\le \frac{k}{n}s) - s\right\} \quad \text{and} \quad \beta_{n,k}(s) = \sqrt{k}\left\{\frac{n}{k}U_{n,[ks]} - s\right\},$$

respectively. In order to derive weighted approximations for the tail empirical process and the quantile process of a dependent sequence, the following regularity conditions are imposed.

**Inference for Heavy-Tailed Data.**
**DOI: http://dx.doi.org/10.1016/B978-0-12-804676-0.00003-1**

**A1)** There exists a sequence $l_n \to \infty$ such that

$$\lim_{n\to\infty}\left\{\frac{\beta(l_n)}{l_n}n + l_n k_n^{-1/2}\log^2(k_n)\right\} = 0,$$

where $\{k_n\}$ satisfies (2.28).

**A2)** There exists a function $r:[0,1]^2 \to \mathbb{R}$ such that for all $0 \le x, y \le 1$

$$\lim_{n\to\infty}\frac{n}{l_n k_n}\mathrm{Cov}\Big(\sum_{i=1}^{l_n} I(U_i > 1-\frac{k_n}{n}x), \sum_{i=1}^{l_n} I(U_i > 1-\frac{k_n}{n}y)\Big) = r(x,y).$$

**A3)** There exists a constant $C > 0$ such that for all $0 \le x < y \le 1$ and $n = 1, 2, \cdots$

$$\frac{n}{l_n k_n}E\left\{\sum_{i=1}^{l_n} I(1-\frac{k_n}{n}y < U_i < 1-\frac{k_n}{n}x)\right\}^4 \le C(y-x).$$

**Theorem 3.1.** *Under conditions of A1)–A3), there exists a centered continuous Gaussian process $W(s)$ with covariance function $r(x,y)$ such that for any $0 \le \nu < 1/2$,*

$$\sup_{0<s<1}\frac{|\alpha_{n,k_n}(s) - W(s)|}{s^\nu} = o_p(1) \quad \textit{and} \quad \sup_{1/(2k_n)\le s\le 1}\frac{|\beta_{n,k_n}(s) + W(s)|}{s^\nu} = o_p(1)$$

*as $n \to \infty$.*

*Proof.* It follows from Theorem 2.3 and Corollary 3.1 of Drees [31]. □

Next we present a weak convergence result for a stationary sequence $\{\xi_i\}_{i=-\infty}^{\infty}$ with a continuous marginal distribution function $F(x)$.

Let $\{\sigma_n > 0\}$ be a sequence of norming constants and let $\{u_n\}$ be a sequence of real numbers. Further let

$$\mathcal{B}_{n,i}^j = \sigma\left\{\xi_i I(\xi_i > u_n), \cdots, \xi_j I(\xi_j > u_n)\right\}$$

be the $\sigma$-field generated by $\xi_i I(\xi_i > u_n), \cdots, \xi_j I(\xi_j > u_n)$. Let $l_n \le r_n \le n$ be sequences of integers and define

$$N_{n,i}(x,y) = \sum_{k=(i-1)r_n+1}^{ir_n} I(u_n + x\sigma_n \le \xi_k \le u_n + y\sigma_n),$$

$$T_n(x) = \frac{\bar{F}(u_n + x\sigma_n)}{\bar{F}(u_n)}, \quad T(x) = (1+\gamma x/\sigma)_+^{-1/\gamma},$$

where $\sigma > 0$ and $\gamma \in \mathbb{R}$. Let $x_T$ denote the right endpoint of the support of $T$, i.e.,

$$x_T = \infty \quad \text{for} \quad \gamma \geq 0 \quad \text{and} \quad x_T = \sigma/|\gamma| \quad \text{for} \quad \gamma < 0.$$

Then the tail empirical distribution function and the tail empirical process are respectively defined as

$$\tilde{T}_n(x) = \frac{1}{n\bar{F}(u_n)} \sum_{i=1}^{n} I(\xi_i > u_n + x\sigma_n)$$
$$\text{and} \quad e(\tilde{T}_n)(x) = \sqrt{n\bar{F}(u_n)} \left\{ \tilde{T}_n(x) - T_n(x) \right\}.$$

To derive the weak convergence, we need the following regularity conditions.

**$\tilde{A}$1)** As $n \to \infty$,

$$\bar{F}(u_n) \to 0, n\bar{F}(u_n) \to \infty, l_n = o(r_n), r_n = o(n), T_n(x) \to T(x) \quad \text{for} \quad x \geq 0.$$

**$\tilde{A}$2)** For each $\theta < x_T$ and for $0 \leq x, y < \theta$, there is a constant $C > 0$ which only depends on $\theta$ such that

$$E\left\{N_{n,1}^p(x, y) | N_{n,1}(x, y) \neq 0\right\} \leq C \quad \text{for some} \quad p \geq 2.$$

**$\tilde{A}$3)** There is a function $r(x, y)$ such that

$$\lim_{n\to\infty} \frac{1}{r_n\bar{F}(u_n)} \text{Cov}\Big(\sum_{i=1}^{r_n} I(\xi_i > u_n + x\sigma_n), \sum_{i=1}^{r_n} I(\xi_i > u_n + y\sigma_n)\Big) = r(x, y).$$

**$\tilde{A}$4)** If $p = 2$ in $\tilde{A}$2), $r_n = o\big((n\bar{F}(u_n))^\nu\big)$ for some $\nu < 1/2$.

**$\tilde{A}$5)** $\beta_n(l_n)n/r_n \to 0$ as $n \to \infty$.

**$\tilde{A}$6)** For some $\mu > 0$,

$$\frac{1}{n\bar{F}(u_n)} \sum_{i,j=1}^{[n/r_n]} \left|\text{Cov}\big(N_{n,i}(x, y), N_{n,j}(x, y)\big)\right| \leq C|x - y|^\mu.$$

**$\tilde{A}$7)** For some $\nu > 0$ and $\theta > 0$

$$\alpha(l_n)n/r_n \to 0, \quad n/r_n = o\big(n\bar{F}^\nu(u_n)\big), \quad \alpha(n) = o(n^{-\theta}).$$

**Theorem 3.2.** *Under either conditions $\tilde{A}1)$–$\tilde{A}5)$ or conditions $\tilde{A}1)$–$\tilde{A}3)$ and $\tilde{A}6)$–$\tilde{A}7)$ with*

$$\frac{p}{p-2} \le \theta, \quad \nu < \frac{\theta}{2} - \frac{\theta+1}{2(p-1)} \quad \text{and} \quad \mu > \frac{2(\theta+p)}{p(\theta+1)},$$

*we have*

$$e(\tilde{T}_n)(x) \xrightarrow{D} e(x) \quad \text{in} \quad D([0, x_T)) \quad \text{as} \quad n \to \infty,$$

*where $e(x)$ is a centered continuous Gaussian process with covariance function $r(x, y)$.*

*Proof.* It follows from Theorems 2.1 and 2.2 of Rootzén [95]. □

## 3.2 HEAVY TAILED DEPENDENT SEQUENCE

Assume $\{X_i\}$ is a strictly stationary sequence with common heavy tailed distribution function $F$ and satisfies the following conditions:

**B1)** $F$ and the sequence $\{k_n\}$ satisfy (2.1) and (2.28), respectively.

**B2)** The sequence $\{X_i\}$ is $\beta$-mixing, i.e., (3.1) holds with $\{U_i\}$ replaced by $\{X_i\}$.

**B3)** There exists a sequence $l_n \to \infty$ such that

$$\lim_{n\to\infty} \left\{ \frac{\beta(l_n)}{l_n} n + l_n k_n^{-1/2} \log^2 k_n \right\} = 0.$$

**B4)** There exist $\epsilon > 0$ and functions $c_m(x, y), m \in \mathcal{N}$ such that for any $m \in \mathcal{N}$ and $0 < x, y \le 1+\epsilon$

$$\lim_{n\to\infty} \frac{n}{k_n} P(F(X_1) > 1 - \frac{k_n}{n}x, F(X_{1+m}) > 1 - \frac{k_n}{n}y)) = c_m(x, y).$$

**B5)** There exist $D_1 \ge 0$ and a sequence $\{\bar{\rho}(m)\}_{m=1}^{\infty}$ such that for any $m \in \mathcal{N}$, $0 < x < y \le 1+\epsilon$ and large $n$

$$\frac{n}{k_n} P\big(X_1 \in I_n(x, y), X_{1+m} \in I_n(x, y)\big) \le (y - x)\big(\bar{\rho}(m) + D_1 \frac{k_n}{n}\big),$$

where $I_n(x, y) = (F^-(1 - yk_n/n), F^-(1 - xk_n/n)]$.

For estimating the tail index $\alpha$ in (2.1) and a high quantile of $F$ based on the dependent observations $X_1, \cdots, X_n$, the corresponding estimators given in Chapter 2 are still applicable, but with a different asymptotic distribution. Here we focus on the Hill estimator

$$\hat{\alpha}(k) = \left\{ \frac{1}{k} \sum_{i=1}^{k} \log \frac{X_{n,n-i+1}}{X_{n,n-k}} \right\}^{-1} \quad \text{given in} \quad (2.31),$$

and the high quantile estimator

$$\hat{x}_p = X_{n,n-k}(np/k)^{-1/\hat{\alpha}(k)} \quad \text{given in} \quad (2.56),$$

where $X_{n,1} \leq \cdots \leq X_{n,n}$ denote the order statistics of $X_1, \cdots, X_n$.

**Theorem 3.3.** *Suppose conditions B1)–B5) and (2.4) hold with*

$$l_n = o(n/k_n) \quad \text{and} \quad \sqrt{k_n} \frac{A(k_n/n)}{(k_n/n)\bar{F}^{-}(k_n/n)} = o(1).$$

*i) We have*

$$\sqrt{k_n}\left\{\hat{\alpha}(k_n) - \alpha\right\} \xrightarrow{d} \alpha \left\{ \int_0^1 t^{-1} W(t)\, dt - W(1) \right\} \quad \text{as} \quad n \to \infty,$$

*where $W(t)$ is a centered Gaussian process with covariance function*

$$c(x, y) = \min(x, y) + \sum_{m=1}^{\infty} \left\{ c_m(x, y) + c_m(y, x) \right\}.$$

*ii) Further assume*

$$p = p_n \to 0, \; k_n^{-1/2} \log(np) \to 0 \text{ and } \frac{np}{k_n} \to 0 \text{ as } n \to \infty.$$

*Then we have*

$$\frac{\sqrt{k_n}}{\log(np/k_n)} \left\{ \frac{\hat{x}_p}{x_p} - 1 \right\} \xrightarrow{d} \alpha \left\{ \int_0^1 t^{-1} W(t)\, dt - W(1) \right\} \quad \text{as} \quad n \to \infty,$$

*where $x_p = \bar{F}^{-}(p)$.*

*Proof.* See Drees [32]. □

Although the above theorem for a strictly stationary sequence is applicable to a time series model, it is quite challenging to verify condition B5) in general. Hence it is useful to develop some alternative efficient methods to infer heavy tailed time series models.

## 3.3 ARMA MODEL

A stationary time series $\{X_t\}$ is said to be an autoregressive moving average model if

$$X_t = \sum_{j=1}^{p} \phi_j X_{t-j} + \varepsilon_t + \sum_{i=1}^{q} \theta_i \varepsilon_{t-i}, \tag{3.2}$$

where $p, q \geq 0$ are integers, $\phi_1, \cdots, \phi_p$ and $\theta_1, \cdots, \theta_q$ are real-valued parameters, and $\{\varepsilon_n\}$ is a sequence of independent and identically distributed random variables satisfying certain moment conditions. We use ARMA$(p, q)$ to denote this model. In particular, an ARMA$(p, 0)$ model is also called an AR$(p)$ model, and an ARMA$(0, q)$ model is also called an MA$(q)$ model.

Consider the following MA$(\infty)$ model:

$$X_t = \sum_{j=0}^{\infty} c_j \varepsilon_{t-j} \quad \text{for} \quad t = 1, \cdots, n, \tag{3.3}$$

where $\{\varepsilon_t\}$ is a sequence of independent and identically distributed random variables with distribution function $G$ satisfying

$$1 - G(x) = rx^{-\alpha} L_1(x) \quad \text{and} \quad G(-x) = (1 - r)x^{-\alpha} L_1(x) \quad \text{for all large} \quad x, \tag{3.4}$$

where $\alpha > 0$, $r \in [0, 1]$ and $L_1(x) \in RV_0^{\infty}$. Further assume that there exist $A > 0$, $u > 1$ and $\delta \in (0, \min(\alpha, 1))$ such that

$$|c_j| \leq Au^{-j} \text{ for large } j, \text{ and } 0 < \sum_{j=0}^{\infty} |c_j|^{\delta} < \infty. \tag{3.5}$$

Therefore

$$\lim_{x \to \infty} \frac{P(|X_1| > x)}{P(|\varepsilon_1| > x)} = \sum_{j=0}^{\infty} |c_j|^{\alpha}, \tag{3.6}$$

i.e., $X_i$ is a heavy tailed random variable with the same tail index $\alpha$ as $\varepsilon_i$. Hence estimating $\alpha$ could be based on either $\{X_i\}_{i=1}^n$ directly or estimated $\{\varepsilon_i\}_{i=1}^n$.

Let $\hat{\alpha}_{|X|}(k)$ denote the Hill estimator based on $\{|X_i|\}_{i=1}^n$, i.e.

$$\hat{\alpha}_{|X|}(k) = \left\{ \frac{1}{k} \sum_{i=1}^{k} \log \frac{|X|_{n,n-i+1}}{|X|_{n,n-k}} \right\}^{-1},$$

where $|X|_{n,1} \leq \cdots \leq |X|_{n,n}$ denote the order statistics of $|X_1|, \cdots, |X_n|$.

**Theorem 3.4.** *Under conditions (3.3), (3.4), (3.5), (2.28),*

$$\sum_{j=1}^{\infty}\sum_{l=0}^{\infty}\min(|c_l|^{\alpha},|c_{j+l}|^{\alpha})\log\frac{\max(|c_l|,|c_{j+l}|)}{\min(|c_l|,|c_{j+l}|)}<\infty,$$

$$\lim_{x\to\infty}xF'(x)/\bar{F}(x)=\alpha \text{ with } F(x)=P(X_1\le x),$$

$$\int_0^{\infty}|G'(x)-G'(x+y)|\,dx<Cy \text{ for some } C>0 \text{ and all } y\in\mathbb{R},$$

$$\text{either } \limsup_{n\to\infty} n/k^{3/2}<\infty \text{ or } \liminf_{n\to\infty} n/k^{3/2}>0,$$

$E|Z_1|^d<\infty$ *for some* $d<1$, *(2.4) and* $\sqrt{k}\frac{A(k/n)}{(k/n)^{1/\alpha}\bar{F}^{-}(k/n)}\to 0$, *we have*

$$\sqrt{k}\left\{\hat{\alpha}_{|X|}(k)-\alpha\right\}\xrightarrow{d}N\Big(0,\alpha^2(1+2\frac{\sum_{j=0}^{\infty}\sum_{l=0}^{\infty}\min(|c_l|^{\alpha},|c_{j+l}|^{\alpha})}{\sum_{j=0}^{\infty}|c_j|^{\alpha}})\Big)\quad as \quad n\to\infty.$$

*Proof.* See Resnick and Stărică [93]. □

Next consider an AR($p$) model:

$$X_t=\sum_{j=1}^{p}\phi_j X_{t-j}+\varepsilon_t \quad \text{for} \quad t=1,\cdots,n, \tag{3.7}$$

where $\{\varepsilon_t\}$ is a sequence of independent and identically distributed random variables with distribution function $G$ satisfying (3.4). Let $\hat{\phi}_i$ denote an estimator of $\phi_i$ for $i=1,\cdots,p$, define residuals $\hat{\varepsilon}_t=X_t-\sum_{i=1}^{p}\hat{\phi}_i X_{t-i}$ for $t=1,\cdots,n$, and let $|\hat{\varepsilon}|_{n,1}\le\cdots\le|\hat{\varepsilon}|_{n,n}$ denote the order statistics of $|\hat{\varepsilon}_1|,\cdots,|\hat{\varepsilon}_n|$. Then the Hill estimator based on $\{|\hat{\varepsilon}_t|\}_{t=1}^{n}$ is defined as

$$\hat{\alpha}_{|\varepsilon|}(k)=\left\{\frac{1}{k}\sum_{i=1}^{k}\log\frac{|\hat{\varepsilon}|_{n,n-i+1}}{|\hat{\varepsilon}|_{n,n-k}}\right\}^{-1}.$$

To derive the asymptotic limit of $\hat{\alpha}_{|\varepsilon|}(k)$ we need the following conditions.

**C1)** All the roots of $1-\phi_1 x\cdots-\phi_p x^p=0$ are outside of the unit circle, which means that $\{X_t\}$ is stationary.

**C2)** $d_n(\hat{\phi}_i-\phi_i)=O_p(1)$ for some $d_n=\sqrt{n}$ when $\alpha>2$ and $d_n=n^{1/\alpha}L(n)$ when $\alpha\in(0,2]$, where $L\in RV_0^{\infty}$.

**C3)** There exists $A(t)\in RV_{\rho}^{0}$ for some $\rho\ge 0$ such that

$$\lim_{t\to 0}\frac{(tx)^{1/\alpha}H^{-}(tx)-t^{1/\alpha}H^{-}(t)}{A(t)}=\frac{x^{\rho}-1}{\rho}\quad \text{for all} \quad x>0,$$

where $H(t)=1-G(t)+G(-t)$.

**C4)** $\lim_{n\to\infty}\sqrt{k}\frac{A(k/n)}{(k/n)^{1/\alpha}H^-(k/n)}=\lambda\in\mathbb{R}$ for the sequence $\{k\}$ in (2.28).

**Theorem 3.5.** *Under conditions (3.7), (3.4) for the distribution of $\varepsilon_t$, C1)–C4), we have*

$$\sqrt{k}\left\{\hat{\alpha}_{|\varepsilon|}(k)-\alpha\right\}\xrightarrow{d}N(\frac{\lambda\alpha^2}{1+\rho},\alpha^2)\quad as\quad n\to\infty.$$

*Proof.* Like the proof of Theorem 2.4, the theorem follows from the following lemma. □

**Remark 3.1.** The above theorem shows that the tail index can be estimated via residuals as if it were estimated based on the true model errors $\{\varepsilon_t\}$. Like Ling and Peng [70], the above estimator based on residuals can be extended to an ARMA model as well.

**Lemma 3.1.** *Put*

$$\hat{W}_n(u)=k^{-1/2}\sum_{j=1}^{n}\left\{I\big(H(|\hat{\varepsilon}_j|)\le\frac{k}{n}u\big)-\frac{k}{n}u\right\}\quad for\quad u\in[0,1].$$

*Then, under conditions of Theorem 3.5, we have*

$$\hat{W}_n(u)\xrightarrow{D}B(u)\quad in\quad D[0,1]\quad as\quad n\to\infty,$$

*where $D[0,1]$ denotes the space of functions on $[0,1]$ which is defined and equipped with the Skorokhod topology given in Chapter 1 and $\{B(u): u>0\}$ is a standard Brownian motion.*

*Proof.* Put

$$\mathbf{X}_{j-1}=(X_{j-1},\cdots,X_{j-p})^T,\quad \boldsymbol{\delta}=(\delta_1,\cdots,\delta_p)^T,$$

$$||\mathbf{X}_{j-1}||=\sum_{i=1}^{p}|X_{j-i}|,\quad h_j(\boldsymbol{\delta},\lambda)=d_n^{-1}\boldsymbol{\delta}^T\mathbf{X}_{j-1}+\lambda d_n^{-1}||\mathbf{X}_{j-1}||,$$

$$\begin{aligned}a_{nj}(u,\boldsymbol{\delta},\lambda)&=I\big(\varepsilon_j\le H^-(\frac{k}{n}u)+h_j(\boldsymbol{\delta},\lambda)\big)-G\big(H^-(\frac{k}{n}u)+h_j(\boldsymbol{\delta},\lambda)\big)\\&\quad+G\big(H^-(\frac{k}{n}u)\big)-I\big(\varepsilon_j\le H^-(\frac{k}{n}u)\big),\end{aligned}$$

and

$$\begin{aligned}b_{nj}(u,\boldsymbol{\delta},\lambda)&=I\big(\varepsilon_j\ge -H^-(\frac{k}{n}u)+h_j(\boldsymbol{\delta},\lambda)\big)-\bar{G}\big(-H^-(\frac{k}{n}u)+h_j(\boldsymbol{\delta},\lambda)\big)\\&\quad+\bar{G}\big(-H^-(\frac{k}{n}u)\big)-I\big(\varepsilon_j\ge -H^-(\frac{k}{n}u)\big).\end{aligned}$$

For any given $M > 0$ such that $\sup_{1\le i\le p}|\delta_i| \le M$ and $|\lambda| \le M$, first we shall show that

$$\sup_{0<u\le 1} |k^{-1/2}\sum_{j=1}^{n} a_{nj}(u, \boldsymbol{\delta}, \lambda)| \xrightarrow{p} 0 \tag{3.8}$$

and

$$\sup_{0<u\le 1} |k^{-1/2}\sum_{j=1}^{n} b_{nj}(u, \boldsymbol{\delta}, \lambda)| \xrightarrow{p} 0. \tag{3.9}$$

Let $\delta' > 0$ be a small number. Define $N = [k^{1/2+\delta'/2}]$ and $u_i = i/N$ for $i = 0, \cdots, N$. When $u \in [u_r, u_{r+1}]$, we have

$$\begin{aligned}
& k^{-1/2}\sum_{j=1}^{n} a_{nj}(u, \boldsymbol{\delta}, \lambda) \\
\le\ & k^{-1/2}\sum_{j=1}^{n} a_{nj}(u_{r+1}, \boldsymbol{\delta}, \lambda) \\
& + k^{-1/2}\sum_{j=1}^{n}\Big\{G\big(H^{-}(\frac{k}{n}u_{r+1}) + h_j(\boldsymbol{\delta}, \lambda)\big) - G\big(H^{-}(\frac{k}{n}u_{r+1})\big)\Big\} \\
& + k^{-1/2}\sum_{j=1}^{n}\Big\{G\big(H^{-}(\frac{k}{n}u_{r})\big) - G\big(H^{-}(\frac{k}{n}u_{r}) + h_j(\boldsymbol{\delta}, \lambda)\big)\Big\} \\
& + 2k^{-1/2}\sum_{j=1}^{n}\Big\{G\big(H^{-}(\frac{k}{n}u_{r+1})\big) - G\big(H^{-}(\frac{k}{n}u_{r})\big)\Big\} \\
& + k^{-1/2}\sum_{j=1}^{n}\Big\{I\big(\varepsilon_j \le H^{-}(\frac{k}{n}u_{r+1})\big) - G\big(H^{-}(\frac{k}{n}u_{r+1})\big) \\
& + G\big(H^{-}(\frac{k}{n}u_{r})\big) - I\big(\varepsilon_j \le H^{-}(\frac{k}{n}u_{r})\big)\Big\}
\end{aligned}$$

and

$$\begin{aligned}
& k^{-1/2}\sum_{j=1}^{n} a_{nj}(u, \boldsymbol{\delta}, \lambda) \\
\ge\ & k^{-1/2}\sum_{j=1}^{n} a_{nj}(u_r, \boldsymbol{\delta}, \lambda) + k^{-1/2}\sum_{j=1}^{n}\Big\{G\big(H^{-}(\frac{k}{n}u_{r}) + h_j(\boldsymbol{\delta}, \lambda)\big) - G\big(H^{-}(\frac{k}{n}u_{r})\big)\Big\} \\
& + k^{-1/2}\sum_{j=1}^{n}\Big\{G\big(H^{-}(\frac{k}{n}u_{r+1})\big) - G\big(H^{-}(\frac{k}{n}u_{r+1}) + h_j(\boldsymbol{\delta}, \lambda)\big)\Big\}
\end{aligned}$$

$$+ 2k^{-1/2}\sum_{j=1}^{n}\Big\{G\big(H^{-}(\frac{k}{n}u_r)\big) - G\big(H^{-}(\frac{k}{n}u_{r+1})\big)\Big\}$$

$$+ k^{-1/2}\sum_{j=1}^{n}\Big\{I\big(\varepsilon_j \le H^{-}(\frac{k}{n}u_r)\big) - G\big(H^{-}(\frac{k}{n}u_r)\big) + G\big(H^{-}(\frac{k}{n}u_{r+1})\big)$$

$$- I\big(\varepsilon_j \le H^{-}(\frac{k}{n}u_{r+1})\big)\Big\}.$$

Hence

$$\sup_{0<u\le 1} |k^{-1/2}\sum_{j=1}^{n} a_{nj}(u, \boldsymbol{\delta}, \lambda)| \le I_1 + I_2 + I_3 + I_4,$$

where

$$I_1 = \sup_r |k^{-1/2}\sum_{j=1}^{n} a_{nj}(u_r, \boldsymbol{\delta}, \lambda)|,$$

$$I_2 = 2\sup_r |k^{-1/2}\sum_{j=1}^{n}\Big\{G\big(H^{-}(\frac{k}{n}u_r) + h_j(\boldsymbol{\delta}, \lambda)\big) - G\big(H^{-}(\frac{k}{n}u_r)\big)\Big\}|,$$

$$I_3 = 2\sup_r |k^{-1/2}\sum_{j=1}^{n}\Big\{G\big(H^{-}(\frac{k}{n}u_{r+1})\big) - G\big(H^{-}(\frac{k}{n}u_r)\big)\Big\}|,$$

and

$$I_4 = \sup_r |k^{-1/2}\sum_{j=1}^{n}\Big\{I\big(\varepsilon_j \le H^{-}(\frac{k}{n}u_r)\big) - G\big(H^{-}(\frac{k}{n}u_r)\big)$$
$$+ G\big(H^{-}(\frac{k}{n}u_{r+1})\big) - I\big(\varepsilon_i \le H^{-}(\frac{k}{n}u_{r+1})\big)\Big\}|.$$

Let $\mathcal{F}_m = \sigma\{\varepsilon_t : t \le m\}$, the $\sigma$-fields generated by $\varepsilon_t$, $1 \le t \le m$, and put

$$s_n = k^{-1/2-\delta'}(k/n)^{-1/\alpha+\delta'/2}.$$

Then for any $\Delta > 0$

$$P(I_1 > \Delta)$$
$$\le N\sup_r P(|k^{-1/2}\sum_{j=1}^{n} a_{nj}(u_r, \boldsymbol{\delta}, \lambda)| > \Delta)$$
$$\le Nk^{-1}\Delta^{-2}\sup_r E(\sum_{j=1}^{n} a_{nj}(u_r, \boldsymbol{\delta}, \lambda))^2$$

$$
\begin{aligned}
&= Nk^{-1}\Delta^{-2}\sup_r \sum_{j=1}^{n} E\left\{E(a_{nj}^2(u_r,\boldsymbol{\delta},\lambda)|\mathcal{F}_{j-1})\right\}\\
&\le Nk^{-1}\Delta^{-2}\sup_r \sum_{j=1}^{n} E|G(H^-(\frac{k}{n}u_r)+h_j(\boldsymbol{\delta},\lambda)) - G(H^-(\frac{k}{n}u_r))|\\
&\le Nk^{-1}\Delta^{-2}\sup_r \sum_{j=1}^{n} E\left\{|G(H^-(\frac{k}{n}u_r)+h_j(\boldsymbol{\delta},\lambda))\right.\\
&\qquad \left. - G(H^-(\frac{k}{n}u_r))|I(d_n^{-1}||\mathbf{X}_{j-1}|| \le s_n)\right\}\\
&\qquad + Nk^{-1}\Delta^{-2}\sup_r \sum_{j=1}^{n} E\left\{|G(H^-(\frac{k}{n}u_r)+h_j(\boldsymbol{\delta},\lambda))\right.\\
&\qquad \left. - G(H^-(\frac{k}{n}u_r))|I(d_n^{-1}||\mathbf{X}_{j-1}|| > s_n)\right\}\\
&= II_1 + II_2.
\end{aligned}
$$

Since

$$
\sup_{0\le u\le 1} |H^-(\frac{k}{n}u)|/s_n \to \infty \text{ and } \sup_{0\le u\le 1} |G'(H^-(\frac{k}{n}u))|/(k/n)^{1+1/\alpha-\delta'/2} \to 0
$$

as $n\to\infty$, we have

$$
\begin{aligned}
II_1 &\le Nk^{-1}\Delta^{-2}\sum_{j=1}^{n} E\left\{|h_j(\boldsymbol{\delta},\lambda)|(k/n)^{1+1/\alpha-\delta'/2} I(d_n^{-1}||\mathbf{X}_{j-1}|| \le s_n)\right\}\\
&\le 2M\Delta^{-2}k^{-\delta'/2} \to 0 \quad \text{as } n\to\infty.
\end{aligned} \tag{3.10}
$$

If $\alpha \le 1$, then by choosing $\delta' > 0$ small enough, we have

$$
\begin{aligned}
II_2 &\le Nk^{-1}\Delta^{-2}2\sum_{j=1}^{n} P(||\mathbf{X}_{j-1}|| > s_n d_n)\\
&\le Nk^{-1}\Delta^{-2}2\sum_{j=1}^{n}\sum_{i=1}^{p} P(|X_{j-i}| > s_n d_n p^{-1})\\
&\le Nk^{-1}\Delta^{-2}2np\{s_n d_n p^{-1}\}^{-\alpha+\delta'} \qquad (3.11)\\
&= 2p^{1+\alpha-\delta'}\Delta^{-2}k^{\delta'\alpha-\delta'\delta'}(k/n)^{-\alpha\delta'/2-\delta'/\alpha+\delta'\delta'/2}k^{1/2+\alpha/2}n^{-1+\delta'/\alpha}(L(n))^{-\alpha+\delta'}\\
&\to 0 \quad \text{as} \quad n\to\infty.
\end{aligned}
$$

If $\alpha > 1$, then we have

$$II_2 \le Nk^{-1}\Delta^{-2}\sup_x |G'(x)|d_n^{-1}\sum_{j=1}^n E\{|\boldsymbol{\delta}^T\mathbf{X}_{j-1}+\lambda||\mathbf{X}_{j-1}|||\times I(||\mathbf{X}_{j-1}||>s_nd_n)\}$$
$$= O(k^{1/2+\delta'/2}k^{-1}nd_n^{-1}(s_nd_n)^{-\alpha+1+\delta'})$$
$$= O(k^{\alpha\delta'-\delta'-\delta'\delta'}(k/n)^{-\alpha\delta'/2+\delta'/2-\delta'/\alpha+\delta'\delta'/2}d_n^{\delta'}(k/n)^{-1/\alpha}d_n^{-\alpha}). \tag{3.12}$$

Note that

$$\begin{cases} k^{\alpha/2}(k/n)^{-1/\alpha}d_n^{-\alpha}=(k/n)^{1-1/\alpha}k^{-1+\alpha/2}(L(n))^{-\alpha} & \text{if } 1<\alpha<2,\\ k^{\alpha/2}(k/n)^{-1/\alpha}d_n^{-\alpha}=(k/n)^{\alpha/2-1/\alpha}(L(n))^{-\alpha} & \text{if } \alpha\ge 2.\end{cases} \tag{3.13}$$

By (3.11)–(3.13), we have $II_2 \to 0$, which implies

$$I_1 \xrightarrow{p} 0 \tag{3.14}$$

by using (3.10).

Write

$$I_2 \le 2k^{-1/2}\sup_r \sum_{j=1}^n |G(H^-(\frac{k}{n}u_r)+h_j(\boldsymbol{\delta},\lambda))-G(H^-(\frac{k}{n}u_r))|$$
$$\times I(d_n^{-1}||\mathbf{X}_{j-1}||\le \bar{s}_n)$$
$$+2k^{-1/2}\sup_r \sum_{j=1}^n |G(H^-(\frac{k}{n}u_r)+h_j(\boldsymbol{\delta},\lambda))-G(H^-(\frac{k}{n}u_r))|$$
$$\times I(d_n^{-1}||\mathbf{X}_{j-1}||>\bar{s}_n)$$
$$=III_1+III_2,$$

where $\bar{s}_n = k^{-1/2-\delta'/2}(k/n)^{-1/\alpha+\delta'/2}$. Similar to the proof of (3.10), we have as $n\to\infty$

$$III_1 \xrightarrow{p} 0 \text{ when } \sup_{1\le i\le p}|\delta_i|\le M,\ |\lambda|\le M. \tag{3.15}$$

If $\alpha \le 1$, by choosing $\delta' > 0$ small enough, we have

$$\sup_{||\boldsymbol{\delta}||\le M}\sup_{|\lambda|\le M} III_2 \le 4k^{-1/2}\sum_{j=1}^n I(d_n^{-1}||\mathbf{X}_{j-1}||>\bar{s}_n)=O_p(k^{-1/2}n(d_n\bar{s}_n)^{-\alpha+\delta'})$$
$$=O_p(k^{\delta'\alpha/2-\delta'/2-\delta'\delta'/2}(k/n)^{-\alpha\delta'/2-\delta'/\alpha+\delta'\delta'/2}k^{1/2+\alpha/2}d_n^{-\alpha+\delta'})=o_p(1).$$

If $\alpha > 1$, then for $\delta' > 0$ small enough

$$
\begin{aligned}
\sup_{||\boldsymbol{\delta}||\le M}\sup_{|\lambda|\le M} III_2 &\le 4Mk^{-1/2}\sup_x |G'(x)| d_n^{-1}\sum_{j=1}^n ||\mathbf{X}_{j-1}|| I(||\mathbf{X}_{j-1}|| > d_n\bar{s}_n) \\
&= O_p(k^{-1/2}d_n^{-1}n(d_n\bar{s}_n)^{-\alpha+1+\delta'}) \\
&= O_p(k^{\alpha\delta'/2-\delta'-\delta'\delta'/2}(k/n)^{-\delta'/\alpha-\alpha\delta'/2+\delta'/2+\delta'\delta'/2}k^{\alpha/2}(k/n)^{-1/\alpha}d_n^{-\alpha+\delta'}) \\
&= o_p(1).
\end{aligned}
$$

Hence by (3.15) we have

$$I_2 \xrightarrow{p} 0. \tag{3.16}$$

It is easy to check that

$$I_3 \xrightarrow{p} 0 \quad \text{and} \quad I_4 \xrightarrow{p} 0. \tag{3.17}$$

Hence (3.8) follows from (3.14), (3.16) and (3.17). Similarly we can derive (3.9).

Put $\hat{\boldsymbol{\delta}}_n = d_n(\hat{\boldsymbol{\phi}} - \boldsymbol{\phi})$,

$$
\begin{aligned}
W_{n,1}(u) &= k^{-1/2}\sum_{j=1}^n \left\{ I\big(\varepsilon_j \le H^-(\frac{k}{n}u)\big) - G\big(H^-(\frac{k}{n}u)\big)\right\}, \\
W_{n,2}(u) &= k^{-1/2}\sum_{j=1}^n \left\{ I\big(\varepsilon_j \ge -H^-(\frac{k}{n}u)\big) - \bar{G}\big(-H^-(\frac{k}{n}u)\big)\right\}, \\
\hat{W}_{n,1}(u) &= k^{-1/2}\sum_{j=1}^n \left\{ I\big(\hat{\varepsilon}_j \le H^-(\frac{k}{n}u)\big) - G\big(H^-(\frac{k}{n}u)\big)\right\},
\end{aligned}
$$

and

$$\hat{W}_{n,2}(u) = k^{-1/2}\sum_{j=1}^n \left\{ I\big(\hat{\varepsilon}_j \ge -H^-(\frac{k}{n}u)\big) - \bar{G}\big(-H^-(\frac{k}{n}u)\big)\right\}.$$

Then

$$
\begin{aligned}
&\hat{W}_{n,1}(u) - W_{n,1}(u) \\
&= k^{-1/2}\sum_{j=1}^n \left\{ G(H^-(\frac{k}{n}u) + \hat{\boldsymbol{\delta}}_n^T d_n^{-1}\mathbf{X}_{j-1}) - G(H^-(\frac{k}{n}u))\right\}
\end{aligned}
$$

$$+ k^{-1/2} \sum_{j=1}^{n} \Big\{ I(\varepsilon_j \le H^{-}(\frac{k}{n}u) + \hat{\boldsymbol{\delta}}_n^T d_n^{-1} \mathbf{X}_{j-1}) - G(H^{-}(\frac{k}{n}u) + \hat{\boldsymbol{\delta}}_n^T d_n^{-1} \mathbf{X}_{j-1})$$
$$+ G(H^{-}(\frac{k}{n}u)) - I(\varepsilon_j \le H^{-}(\frac{k}{n}u)) \Big\}.$$

Define $D_\Delta = [-\Delta, \Delta]^p$ for some $\Delta > 0$,

$$E_{n1}(u, \boldsymbol{\delta}) = k^{-1/2} \sum_{j=1}^{n} \left\{ G(H^{-}(\frac{k}{n}u)) + d_n^{-1} \boldsymbol{\delta}^T \mathbf{X}_{j-1}) - \frac{k}{n}u \right\},$$

and

$$E_{n2}(u, \boldsymbol{\delta}) = k^{-1/2} \sum_{j=1}^{n} \Big\{ I(\varepsilon_j \le H^{-}(\frac{k}{n}u) + d_n^{-1} \boldsymbol{\delta}^T \mathbf{X}_{j-1})$$
$$- G(H^{-}(\frac{k}{n}u) + d_n^{-1} \boldsymbol{\delta}^T \mathbf{X}_{j-1}) + G(H^{-}(\frac{k}{n}u)) - I(\epsilon_j \le H^{-}(\frac{k}{n}u)) \Big\}.$$

Next we show that

$$\sup_{\boldsymbol{\delta} \in D_\Delta} \sup_{0 \le u \le 1} |E_{n1}(u, \boldsymbol{\delta})| = o_p(1) \tag{3.18}$$

and

$$\sup_{\boldsymbol{\delta} \in D_\Delta} \sup_{0 \le u \le 1} |E_{n2}(u, \boldsymbol{\delta})| = o_p(1). \tag{3.19}$$

Let $\boldsymbol{\delta}_r$ be a fixed point in $D_\Delta$. Then for any $\boldsymbol{\delta} \in D_\Delta$, we have

$$|h_j(\boldsymbol{\delta}, \lambda) - h_j(\boldsymbol{\delta}_r, \lambda)| \le ||\boldsymbol{\delta} - \boldsymbol{\delta}_r|| |d_n^{-1}| ||\mathbf{X}_{j-1}|| \le p\Delta d_n^{-1} ||\mathbf{X}_{j-1}||,$$

i.e.,

$$h_j(\boldsymbol{\delta}_r, \lambda - p\Delta) \le h_j(\boldsymbol{\delta}, \lambda) \le h_j(\boldsymbol{\delta}_r, \lambda + p\Delta),$$

which implies that

$$E_{n2}(u, \boldsymbol{\delta}) \le \frac{1}{\sqrt{k}} \sum_{j=1}^{n} a_{nj}(u, \boldsymbol{\delta}_r, p\Delta)$$
$$+ \frac{1}{\sqrt{k}} \sum_{j=1}^{n} \left\{ G(H^{-}(\frac{k}{n}u) + h_j(\boldsymbol{\delta}_r, p\Delta)) - G(H^{-}(\frac{k}{n}u) + h_j(\boldsymbol{\delta}, 0)) \right\} \tag{3.20}$$

and

$$E_{n2}(u,\boldsymbol{\delta}) \geq \frac{1}{\sqrt{k}}\sum_{j=1}^{n} a_{nj}(u,\boldsymbol{\delta}_r,-p\Delta) + \frac{1}{\sqrt{k}}\sum_{j=1}^{n}\left\{G(H^-(\frac{k}{n}u)+h_j(\boldsymbol{\delta}_r,-p\Delta)) - G(H^-(\frac{k}{n}u)+h_j(\boldsymbol{\delta},0))\right\}. \tag{3.21}$$

Similar to the proof of (3.16), we have

$$\sup_{0\leq u\leq 1}\sup_{\boldsymbol{\delta}\in D_\Delta}\frac{1}{\sqrt{k}}\left|\sum_{j=1}^{b}\left\{G\big(H^-(\frac{k}{n}u)+h_j(\boldsymbol{\delta}_r,\pm p\Delta)\big) - G\big(H^-(\frac{k}{n}u)+h_j(\boldsymbol{\delta},0)\big)\right\}\right| \xrightarrow{p} 0. \tag{3.22}$$

Then it follows from (3.20) and (3.22) that for any $\epsilon > 0$,

$$\begin{aligned}
&P\big(\sup_{\boldsymbol{\delta}\in D_\Delta}\sup_{0\leq u\leq 1}|E_{n2}(u,\boldsymbol{\delta})|\geq\epsilon\big)\\
\leq\ & P\big(\sup_{0\leq u\leq 1}\big|\frac{1}{\sqrt{k}}\sum_{j=1}^{n}a_{nj}(u,\boldsymbol{\delta}_r,p\Delta)\big|\geq\epsilon/4\big)\\
&+P\big(\sup_{0\leq u\leq 1}\big|\frac{1}{\sqrt{k}}\sum_{j=1}^{n}a_{nj}(u,\boldsymbol{\delta}_r,-p\Delta)\big|\geq\epsilon/4\big)\\
&+P\big(\sup_{\boldsymbol{\delta}\in D_\Delta}\sup_{0\leq u\leq 1}\big|\frac{1}{\sqrt{k}}\sum_{j=1}^{n}\{G(H^-(\frac{k}{n}u)+h_j(\boldsymbol{\delta}_r,p\Delta))\\
&-G(H^-(\frac{k}{n}u)+h_j(\boldsymbol{\delta},0))\}\big|\geq\epsilon/4\big)\\
&+P\big(\sup_{\boldsymbol{\delta}\in D_\Delta}\sup_{0\leq u\leq 1}\big|\frac{1}{\sqrt{k}}\sum_{j=1}^{n}\{G(H^-(\frac{k}{n}u)+h_j(\boldsymbol{\delta}_r,-p\Delta))\\
&-G(H^-(\frac{k}{n}u)+h_j(\boldsymbol{\delta},0))\}\big|\geq\epsilon/4\big)\\
\to\ & 0,
\end{aligned}$$

i.e., (3.19) holds. Obviously (3.18) follows from (3.16). Therefore

$$\sup_{0\leq u\leq 1}|\hat{W}_{n,1}(u)-W_{n,1}(u)|\xrightarrow{p}0.$$

Similarly we have

$$\sup_{0 \le u \le 1} |\hat{W}_{n,2}(u) - W_{n,2}(u)| \xrightarrow{p} 0.$$

Hence the theorem follows from the facts that both $\hat{W}_n(u) = \hat{W}_{n,1}(u) + \hat{W}_{n,2}(u)$ and $W_{n,1}(u) + W_{n,2}(u)$ converge weakly to $B(u)$ in the space $D[0,1]$. □

## 3.4 STOCHASTIC DIFFERENCE EQUATIONS

A random first order difference equation is

$$R_{n+1} = Q_{n+1} + M_{n+1} R_n, \quad n = 0, 1, \cdots,$$

where $\{(Q_n, M_n)^T : n = 1, 2, \cdots\}$ are i.i.d. random vectors with some given joint distribution function and $R_0$ is independent of these with some given starting distribution function. If the joint distribution function of $Q_n$ and $M_n$ (denoted by $(Q, M)^T$) satisfies appropriate conditions, the distribution function of $R_n$ will converge to a limit that does not depend on $R_0$ and be the unique solution to the random functional equation

$$R \stackrel{d}{=} Q + MR, \quad \text{where } R \text{ is independent of } (Q, M)^T.$$

The following theorem describes the tail behavior of the distributions of $R$ and $Q$.

**Theorem 3.6.** *Let* $(Q, M)^T$ *be a random vector with*

$$E(\log^+ |Q|) < \infty, \; P(M \ge 0) = 1, \; E(M^\alpha) < 1 \; \text{ and } E(M^\beta) < \infty$$

*for some* $\beta > \alpha > 0$. *Let* $R$ *be a random variable independent of* $Q$ *and* $M$. *Then there exists exactly one distribution function for* $R$ *such that* $Q + MR$ *has the same distribution function as* $R$. *If* $R$ *has this distribution function and* $L \in RV_0^\infty$, *then as* $t \to \infty$,

$$P(Q > t) = t^{-\alpha} L(t) \;\text{ is equivalent to } P(R > t) = \frac{1}{1 - E(M^\alpha)} t^{-\alpha} L(t).$$

*Proof.* See Theorem 1 of Grey [46]. □

## 3.5 HEAVY TAILED GARCH SEQUENCES

A generalized autoregressive conditionally heteroscedastic process $\{X_t\}$ of order $(p, q)$ with $p, q \geq 0$ (called GARCH$(p, q)$) is given by

$$X_t = \sigma_t \varepsilon_t, \quad \sigma_t^2 = \alpha_0 + \sum_{i=1}^{p} \alpha_i X_{t-i}^2 + \sum_{j=1}^{q} \beta_j \sigma_{t-j}^2, \tag{3.23}$$

where $\{\varepsilon_t\}$ is a sequence of independent and identically distributed random variables with mean zero and variance one.

Define the $(p+q-1) \times (p+q-1)$ matrix

$$\mathbf{A}_t = \begin{pmatrix} \alpha_1 \varepsilon_t^2 + \beta_1 & \beta_2 & \dots & \beta_{q-1} & \beta_q & \alpha_2 & \alpha_3 & \dots & \alpha_p \\ 1 & 0 & \dots & 0 & 0 & 0 & 0 & \dots & 0 \\ 0 & 1 & \dots & 0 & 0 & 0 & 0 & \dots & 0 \\ \vdots & \vdots & \ddots & \vdots & \vdots & \vdots & \vdots & \ddots & \vdots \\ 0 & 0 & \dots & 1 & 0 & 0 & 0 & \dots & 0 \\ \varepsilon_t^2 & 0 & \dots & 0 & 0 & 0 & 0 & \dots & 0 \\ 0 & 0 & \dots & 0 & 0 & 1 & 0 & \dots & 0 \\ \vdots & \vdots & \ddots & \vdots & \vdots & \vdots & \vdots & \ddots & \vdots \\ 0 & 0 & \dots & 0 & 0 & 0 & \dots & 1 & 0 \end{pmatrix}$$

and the $(p+q-1) \times 1$ matrix $\mathbf{B}_t = (\alpha_0, 0, \dots, 0)^T$. Denote the Euclidean $L_2$-norm in $\mathbb{R}^{p+q}$ by $|\cdot|_2$ as before and the operator norm for matrix $\mathbf{A}_t$ by $\|\mathbf{A}_t\| = \sup_{|x|_2=1} |\mathbf{A}_t x|_2$. Then the Lyapunov exponent for the sequence of random matrices $\{\mathbf{A}_t\}$ is given by

$$\gamma = \inf \left\{ \frac{1}{n} E(\log \|\mathbf{A}_1 \cdots \mathbf{A}_n\|) : \ n = 1, 2, \cdots \right\}.$$

Put

$$\mathbf{X}_t = (X_{t,1}, \cdots, X_{t,p+q-1})^T := (\sigma_{t+1}^2, \cdots, \sigma_{t-q+2}^2, X_t^2, \cdots, X_{t-p+2}^2)^T.$$

Some basic properties of a GARCH sequence are summarized in the following theorem.

**Theorem 3.7.** *Suppose (3.23) holds with $\alpha_0 > 0$ and $\gamma < 0$.*

**i)** *If $E \log^+ |\varepsilon_1| < \infty$, then there exists a unique strictly stationary causal solution of (3.23).*

**ii)** *If* $\varepsilon_1$ *has a positive density function on* $\mathbb{R}$ *such that* $E|\varepsilon_1|^h < \infty$ *for all* $h < h_0$ *and* $E|\varepsilon_1|^{h_0} = \infty$ *for some* $h_0 \in (0, \infty]$, *and not all of* $\alpha_1, \cdots, \alpha_p, \beta_1, \cdots, \beta_q$ *vanish, then there exist* $\kappa > 0$ *and a finite-valued function* $w(\boldsymbol{x})$ *such that*

$$\lim_{u\to\infty} u^\kappa P(\boldsymbol{x}^T \boldsymbol{X}_1 > u) = w(\boldsymbol{x}) \quad \text{for any} \quad \boldsymbol{x} \neq \boldsymbol{0}.$$

**iii)** *If* $\varepsilon_1$ *has a density function positive in an interval containing zero, then* $\{\boldsymbol{X}_t\}$ *is strongly mixing with the geometric rate.*

*Proof.* See Theorem 3.1 of Basrak et al. [4]. □

Although the Hill estimator in (2.31) may be applicable to estimating the tail index of $X_t$, checking conditions in Theorem 3.3 is very difficult for a GARCH$(p, q)$ process. However, for a GARCH$(1, 1)$ sequence, the tail index can be estimated via an estimating equation, which leads to a much more efficient estimator than the Hill estimator. Moreover, many financial data sets can be fitted well by a GARCH$(1, 1)$ model or an ARMA–GARCH$(1, 1)$ model.

Consider a GARCH$(1, 1)$ model:

$$Y_t = \sigma_t^* \varepsilon_t^*, \ (\sigma_t^*)^2 = \omega^* + a^* Y_{t-1}^2 + b^*(\sigma_{t-1}^*)^2, \tag{3.24}$$

where $\omega^* > 0$, $a^* \geq 0$, $b^* \geq 0$, and $\{\varepsilon_t^*\}$ is a sequence of independent and identically distributed random variables with zero mean and unit variance. Under conditions in Theorem 3.7, we have for some $c > 0$

$$P(|Y_t| > x) = cx^{-\alpha}\{1 + o(1)\} \text{ as } x \to \infty, \tag{3.25}$$

and the tail index $\alpha$ is determined by

$$E\{b^* + a^*(\varepsilon_t^*)^2\}^{\alpha/2} = 1; \tag{3.26}$$

see Mikosch and Stărică [74] for more details.

When $E|\varepsilon_t^*|^\delta < \infty$ for some $\delta > \max\{4, 2\alpha\}$, one can first estimate the nuisance parameters $\boldsymbol{\theta}^* = (\omega^*, a^*, b^*)^T$ by the quasi-maximum likelihood estimator (QMLE), that is,

$$\hat{\boldsymbol{\theta}}^* = (\hat{\omega}^*, \hat{a}^*, \hat{b}^*)^T = \arg\max_{\boldsymbol{\theta}^*} \sum_{t=1}^n \log\left\{\frac{1}{\sqrt{2\pi}\sigma_t^*} \exp\{-\frac{Y_t^2}{2\sigma_t^{*2}}\}\right\}.$$

Using the obtained QMLE and Eq. (3.26), the tail index $\alpha$ can be estimated via solving the following estimating equation:

$$\frac{1}{n}\sum_{t=1}^{n}\{\hat{b}^* + \hat{a}^*(\hat{\varepsilon}_t^*)^2\}^{\alpha/2} = 1,$$

where $\hat{\varepsilon}_t^* = Y_t/\hat{\sigma}_t^*$ and $\hat{\sigma}_t^*$ is an estimator of $\sigma_t^*$ with $\boldsymbol{\theta}^*$ being replaced by $\hat{\boldsymbol{\theta}}^*$, see Berkes et al. [8] for the asymptotic distribution of the above estimator and Chan et al. [17] for a profile empirical likelihood inference based on the above estimation procedure.

It follows from Hall and Yao [54] that the QMLE $\hat{\boldsymbol{\theta}}^*$ has a nonnormal limit when $E|\varepsilon_t^*|^4 = \infty$. To relax the moment condition on $\varepsilon_t^*$, Zhang et al. [106] proposed to employ the least absolute deviations estimator (LADE) in Peng and Yao [85] as follows.

Assume the unknown median of $(\varepsilon_t^*)^2$ is $d > 0$ and put $\varepsilon_t = \varepsilon_t^*/\sqrt{d}$. Then the median of $\log \varepsilon_t^2$ becomes $\log\{\text{median}((\varepsilon_t^*)^2/d)\} = 0$. Furthermore, model (3.24) and Eq. (3.26) can be written as

$$Y_t = \sigma_t \varepsilon_t, \quad \sigma_t^2 = \omega + aY_{t-1}^2 + b\sigma_{t-1}^2 \tag{3.27}$$

and

$$E\{b + a\varepsilon_t^2\}^{\alpha/2} = 1, \tag{3.28}$$

where $\sigma_t = \sqrt{d}\sigma_t^*$, $\omega = d\omega^*$, $a = da^*$ and $b = b^*$.

For $\boldsymbol{\theta} = (\omega, a, b)^T$, by the recursion of (3.27), the conditional variance $\sigma_t^2 = \sigma_t^2(\boldsymbol{\theta})$ can be represented as

$$\sigma_t^2(\boldsymbol{\theta}) = \omega + aY_{t-1}^2 + b\sigma_{t-1}^2(\theta) = \frac{\omega(1-b^t)}{1-b} + \sum_{k=0}^{t-1} ab^k Y_{t-1-k}^2 + b^t\sigma_0^2(\boldsymbol{\theta}), \tag{3.29}$$

and a truncated version is

$$\bar{\sigma}_t^2(\boldsymbol{\theta}) = \omega(1-b^t)/(1-b) + a\sum_{0\le k\le t-1} b^k Y_{t-k-1}^2.$$

Then the least absolute deviations estimator (LADE) in Peng and Yao [85] is

$$\hat{\boldsymbol{\theta}} = \arg\min_{\boldsymbol{\theta}} \sum_{t=1}^{n} |\log Y_t^2 - \log \bar{\sigma}_t^2(\boldsymbol{\theta})|. \tag{3.30}$$

As before, using this LADE $\hat{\boldsymbol{\theta}}$, $\alpha$ can be estimated by solving

$$\frac{1}{n}\sum_{t=1}^{n}\left\{\hat{b}+\hat{a}\bar{\varepsilon}_t^2(\hat{\boldsymbol{\theta}})\right\}^{\alpha/2}=1,$$

where $\bar{\varepsilon}_t^2(\hat{\boldsymbol{\theta}})=Y_t^2/\bar{\sigma}_t^2(\hat{\boldsymbol{\theta}})$. Denote this estimator by $\hat{\alpha}_{garch}$. For deriving the asymptotic distribution of $\hat{\alpha}_{garch}$, we need some regularity conditions.

**D1)** $E\log(b_0^*+a_0^*(\varepsilon_t^*)^2)<0$ (i.e., $E\log(b_0+a_0\varepsilon_t^2)<0$) and $E|\varepsilon_t^*|^{\delta_0}<\infty$ (i.e., $E|\varepsilon_t|^{\delta_0}<\infty$) for some $\delta_0>\max\{2,2\alpha_0\}$, where $\boldsymbol{\theta}_0=(\omega_0,a_0,b_0)^T$, $\boldsymbol{\theta}_0^*=(\omega_0^*,a_0^*,b_0^*)^T$ and $\alpha_0$ denote the true values of $\boldsymbol{\theta}$, $\boldsymbol{\theta}^*$ and $\alpha$ respectively.

**D2)** $(\varepsilon_t^*)^2$ has an unknown median $d>0$ and a continuous density function at $d$, i.e., $\log\{\varepsilon_t^2\}$ has median zero and its density function $f(x)$ is continuous at zero.

**Theorem 3.8.** *Assume conditions D1) and D2) hold for model (3.24) with Eq. (3.26). Then, as $n\to\infty$*

$$\sqrt{n}\left\{\hat{\alpha}_{garch}-\alpha_0\right\}\xrightarrow{d}N(0,\gamma_{\alpha_0}^2),$$

*where*

$$\begin{aligned}\gamma_{\alpha_0}^2&=\{4A_0^2f^2(0)\}^{-1}(\mu_1,\mu_2,\mu_3)\Omega^{-1}(\mu_1,\mu_2,\mu_3)^T\\&\quad+4\{A_0^2\}^{-1}E\big((b_0+a_0\varepsilon_1^2)^{\frac{\alpha_0}{2}}-1\big)^2\\&\quad+2\{A_0^2f(0)\}^{-1}(\mu_1,\mu_2,\mu_3)\Omega^{-1}E\{A(1)\big((b_0+a_0\varepsilon_1^2)^{\frac{\alpha_0}{2}}-1\big)\}\end{aligned}$$

*with*

$$A_0=E\big((b_0+a_0\varepsilon_1^2)^{\frac{\alpha_0}{2}}\log(b_0+a_0\varepsilon_1^2)\big),\qquad e_0=\alpha_0E\big((b_0+a_0\varepsilon_1^2)^{\alpha_0/2-1}\varepsilon_1^2\big),$$

$$\mu_1=-\frac{a_0e_0}{2}E\frac{\partial\log\sigma_1^2(\boldsymbol{\theta}_0)}{\partial w},\quad\mu_2=\alpha_0E\big((b_0+a_0\varepsilon_1^2)^{\frac{\alpha_0}{2}-1}\big)-\frac{a_0e_0}{2}E\frac{\partial\log\sigma_1^2(\boldsymbol{\theta}_0)}{\partial b},$$

$$\mu_3=e_0-\frac{a_0e_0}{2}E\frac{\partial\log\sigma_1^2(\boldsymbol{\theta}_0)}{\partial a},\qquad\Omega=E\big(A(1)A^T(1)\big),$$

$$A(t)=\Big(\frac{\partial(\log\sigma_t^2(\boldsymbol{\theta}_0))}{\partial\omega},\frac{\partial(\log\sigma_t^2(\boldsymbol{\theta}_0))}{\partial b},\frac{\partial(\log\sigma_t^2(\boldsymbol{\theta}_0))}{\partial a}\Big)^T sgn\{\log(\varepsilon_t^2)\},$$

*and sgn denotes the sign function.*

*Proof.* See Zhang et al. [106]. □

The above theorem shows that the proposed tail index estimator for a GARCH(1, 1) sequence has the standard rate of convergence $n^{-1/2}$, which is faster than the rate $k^{-1/2}$ of the Hill estimator in Theorem 3.3.

Although an interval for $\alpha$ can be obtained via either estimating the asymptotic variance of $\hat{\alpha}_{garch}$ or using a bootstrap method, an alternative method is to employ an empirical likelihood method which has been shown to be efficient in interval estimation and hypothesis tests in many situations.

Since the proposed LADE is a solution to the score equations

$$\sum_{t=1}^{n} \bar{Z}_{t,j}(\boldsymbol{\theta}) = 0 \quad \text{for} \quad j = 2, 3, 4,$$

where

$$\begin{aligned}
\bar{Z}_{t,2}(\boldsymbol{\theta}) &= \left(\partial(\log\bar{\sigma}_t^2(\boldsymbol{\theta}))/\partial\omega\right)\operatorname{sgn}\{\log(Y_t^2/\bar{\sigma}_t^2(\boldsymbol{\theta}))\},\\
\bar{Z}_{t,3}(\boldsymbol{\theta}) &= \left(\partial(\log\bar{\sigma}_t^2(\boldsymbol{\theta}))/\partial a\right)\operatorname{sgn}\{\log(Y_t^2/\bar{\sigma}_t^2(\boldsymbol{\theta}))\},\\
\bar{Z}_{t,4}(\boldsymbol{\theta}) &= \left(\partial(\log\bar{\sigma}_t^2(\boldsymbol{\theta}))/\partial b\right)\operatorname{sgn}\{\log(Y_t^2/\bar{\sigma}_t^2(\boldsymbol{\theta}))\},
\end{aligned}$$

we know that $\boldsymbol{\theta}$ and $\alpha$ can be estimated simultaneously by solving the following equations

$$\sum_{t=1}^{n} \bar{Z}_{t,1}(\boldsymbol{\theta},\alpha) = 0 \quad \text{and} \quad \sum_{t=1}^{n} \bar{Z}_{t,j}(\boldsymbol{\theta}) = 0 \quad \text{for} \quad j = 2, 3, 4,$$

where

$$\bar{Z}_{t,1}(\boldsymbol{\theta},\alpha) = \{b + aY_t^2/\bar{\sigma}_t^2(\boldsymbol{\theta})\}^{\alpha/2} - 1.$$

This motivates the application of the so-called profile empirical likelihood method based on estimating equations in Qin and Lawless [89].

Put

$$\bar{\boldsymbol{Z}}_t(\boldsymbol{\theta},\alpha) = (\bar{Z}_{t,1}(\boldsymbol{\theta},\alpha), \bar{Z}_{t,2}(\boldsymbol{\theta}), \bar{Z}_{t,3}(\boldsymbol{\theta}), \bar{Z}_{t,4}(\boldsymbol{\theta}))^T \quad \text{for} \quad t = 1,\ldots,n,$$

and define the empirical likelihood function of $\boldsymbol{\theta}$ and $\alpha$ as

$$L(\boldsymbol{\theta},\alpha) = \sup\left\{\prod_{t=1}^{n}(np_t) : p_1 \geq 0,\ldots,p_n \geq 0, \sum_{t=1}^{n} p_t = 1, \sum_{t=1}^{n} p_t\bar{\boldsymbol{Z}}_t(\boldsymbol{\theta},\alpha) = 0\right\}.$$

The following theorem comes from Zhang et al. [106], which says that Wilks theorem holds for the proposed empirical likelihood method and an empirical likelihood confidence interval with level $\xi$ is obtained as

$$I_\xi = \left\{ \alpha : -2\log\left(\max_{\boldsymbol{\theta}} L(\boldsymbol{\theta}, \alpha)\right) \le \chi^2_{1,\xi} \right\},$$

where $\chi^2_{1,\xi}$ is the $\xi$-quantile of a chi-squared distribution with one degree of freedom.

**Theorem 3.9.** *Under conditions of Theorem 3.8,* $-2\log\left(\max_{\boldsymbol{\theta}} L(\boldsymbol{\theta}, \alpha_0)\right)$ *converges in distribution to a chi-squared distribution with one degree of freedom as* $n \to \infty$.

*Proof.* See Zhang et al. [106]. □

## 3.6 DOUBLE AR(1) MODEL

This subsection studies another time series model called the first-order autoregressive model (AR(1)) with autoregressive conditional heteroscedastic errors of order one (ARCH(1)):

$$Y_t = a^* Y_{t-1} + \sqrt{\omega^* + b^* Y_{t-1}^2}\,\varepsilon_t^*, \tag{3.31}$$

where $\{\varepsilon_t^*\}$ is a sequence of independent and identically distributed random variables with zero mean and unit variance, $a^* \in \mathbb{R}$, $\omega^* > 0$ and $b^* > 0$. This model is also called a double AR model in the literature. Throughout assume the following regularity conditions.

**E1)** $E\log(|a^* + \sqrt{b^*}\varepsilon_1^*|) < 0$.

**E2)** $\varepsilon_t^*$ has a symmetric, positive and continuous Lebesgue density function on $\mathbb{R}$.

Then it follows from Borkovec and Klüppelberg [10] that $Y_t$ has a heavy tail with index $\alpha > 0$, which is determined by

$$E(|a^* + \sqrt{b^*}\varepsilon_t^*|^\alpha) = 1. \tag{3.32}$$

Although one may apply the Hill estimator in (2.31) to $Y_t$'s directly and derive the asymptotic distribution by using Theorem 3.3, checking regularity conditions in Theorem 3.3 is nontrivial at all. Moreover the Hill estimator has a rate of convergence slower than $n^{-1/2}$. As before, one could estimate $\alpha$ by solving

$$\frac{1}{n}\sum_{t=1}^{n} |\hat{a}^* + \sqrt{\hat{b}^*}\hat{\varepsilon}_t^*|^\alpha = 1,$$

where $\hat{a}^*, \hat{b}^*, \hat{\varepsilon}_t^*$ are some estimators for $a^*, b^*, \varepsilon_t^*$, respectively. Indeed Chan et al. [18] proposed to first employ the QMLE in Ling [68] to estimate $\alpha$ and then to apply a profile empirical likelihood method for interval estimation, where finite fourth moment of $\varepsilon_t^*$ is required to ensure a normal limit. In order to relax this moment condition, Zhang et al. [106] proposed to use the weighted least absolute deviations estimator in Chan and Peng [13] as follows.

Assume the unknown median of $(\varepsilon_t^*)^2$ is $d > 0$. Put $\varepsilon_t = \varepsilon_t^*/\sqrt{d}$. Then the median of $\varepsilon_t^2$ becomes one, and model (3.31) and Eq. (3.32) can be written as

$$Y_t = aY_{t-1} + \sqrt{\omega + bY_{t-1}^2}\,\varepsilon_t \tag{3.33}$$

and

$$E\{|a + \sqrt{b}\epsilon_t|^\alpha\} = 1, \tag{3.34}$$

where $a = a^*, \omega = d\omega^*$ and $b = db^*$. As before, $\boldsymbol{\theta} = (\omega, a, b)^T$ can be estimated by the following weighted least absolute deviations estimator

$$\hat{\boldsymbol{\theta}} = (\hat{\omega}, \hat{a}, \hat{b})^T = \arg\min_{\theta} \sum_{t=1}^{n} \frac{1}{1 + Y_{t-1}^2} |(Y_t - aY_{t-1})^2 - (\omega + bY_{t-1}^2)|. \tag{3.35}$$

After computing $\hat{\varepsilon}_t = (Y_t - \hat{a}Y_{t-1})/\sqrt{\hat{\omega} + \hat{b}Y_{t-1}^2}$, $\alpha$ is estimated by solving the following equation:

$$\frac{1}{n}\sum_{t=1}^{n} |\hat{a} + \sqrt{\hat{b}}\hat{\varepsilon}_t|^\alpha = 1.$$

Denote this estimator by $\hat{\alpha}_{dar}$, and let $\alpha_0$ denote the true value of $\alpha$.

Put $\Delta = (1 + Y_1^2)(\omega_0 + b_0 Y_1^2)$, $S = 1 + Y_1^2$,

$$\Gamma_1 = \begin{pmatrix} E\frac{a_0^2 Y_1^4}{\Delta} + E\frac{Y_1^2}{S} & E\frac{a_0 Y_1^2}{\Delta} & -E\frac{a_0 Y_1^4}{\Delta} \\ E\frac{a_0 Y_1^2}{\Delta} & E\frac{1}{\Delta} & -E\frac{Y_1^2}{\Delta} \\ -E\frac{a_0 Y_1^4}{\Delta} & -E\frac{Y_1^2}{\Delta} & E\frac{Y_1^4}{\Delta} \end{pmatrix}, \quad \Gamma_2 = \begin{pmatrix} \frac{1}{2} & 0 & 0 \\ 0 & 1 & 0 \\ \alpha_0 & 0 & 1 \end{pmatrix}.$$

Let $\bar{A}(t) = (Y_t Y_{t-1}, 1, -Y_{t-1}^2)^T \text{sgn}(\varepsilon_t^2 - 1)/(1 + Y_{t-1}^2)$, $f(x)$ denote the density function of $\varepsilon_1$,

$$\bar{\gamma}_{\alpha_0}^2 = \{f(1)\}^{-2}(c_1, c_2, c_3)\Gamma_2\Gamma_1^{-1}\text{Cov}\{\bar{A}(1)\}\Gamma_1^{-1}\Gamma_2(c_1, c_2, c_3)^T$$

$$+\kappa_0^{-2}\mathrm{Var}(|a_0+\sqrt{b_0}\varepsilon_1|^{\alpha_0})-2\{f(1)\}^{-1}\kappa_0^{-1}(c_1,c_2,c_3)$$
$$\times\Gamma_2\Gamma_1^{-1}E\{\bar{A}(1)(|a_0+\sqrt{b_0}\varepsilon_1|^{\alpha_0}-E|a_0+\sqrt{b_0}\varepsilon_1|^{\alpha_0})\},$$

where

$$\kappa_0=\mathrm{E}\{|a_0+\sqrt{b_0}\varepsilon_1|^{\alpha_0}\log|a_0+\sqrt{b_0}\varepsilon_1|\},$$
$$c_1=\kappa_0^{-1}\mathrm{E}\frac{\sqrt{b_0}(\alpha_0|a_0+\sqrt{b_0}\varepsilon_2|^{\alpha_0-1}\mathrm{sgn}(a_0+\sqrt{b_0}\varepsilon_2))\epsilon_2}{2(w_0+b_0Y_1^2)},$$
$$c_2=\kappa_0^{-1}\mathrm{E}\{(\alpha_0|a_0+\sqrt{b_0}\varepsilon_2|^{\alpha_0-1}\mathrm{sgn}(a_0+\sqrt{b_0}\varepsilon_2))(\frac{\sqrt{b_0}Y_1}{\sqrt{w_0+b_0Y_1^2}}-1)\},$$

and

$$c_3=\kappa_0^{-1}\mathrm{E}\{(\alpha_0|a_0+\sqrt{b_0}\varepsilon_2|^{\alpha_0-1}\mathrm{sgn}(a_0+\sqrt{b_0}\varepsilon_2))(\frac{\sqrt{b_0}\varepsilon_2Y_1^2}{2(w_0+b_0Y_1^2)}-\frac{\varepsilon_2}{2\sqrt{b_0}})\}.$$

The asymptotic distribution of the above proposed tail index estimator is given below.

**Theorem 3.10.** *In addition to conditions E1) and E2) for model (3.31), we further assume that* $\alpha_0>1$ *and* $\mathrm{E}|\varepsilon_t|^{\delta_0}<\infty$ *for some* $\delta_0>2\alpha_0$. *Then*

$$\sqrt{n}\left\{\hat{\alpha}_{dar}-\alpha_0\right\}\xrightarrow{d}\mathrm{N}(0,\bar{\gamma}_{\alpha_0}^2)\quad as\quad n\to\infty.$$

*Proof.* See Zhang et al. [106]. □

Again, to avoid estimating $\bar{\gamma}_{\alpha_0}^2$, a profile empirical likelihood method can be developed to construct a confidence interval for $\alpha_0$. Put

$$Z_t(\omega,a,b)=\left\{(Y_t-aY_{t-1})^2-(\omega+bY_{t-1}^2)\right\}/(1+Y_{t-1}^2),$$

and define

$$X_{t,1}(\boldsymbol{\theta},\alpha)=\left|a+\sqrt{b}(Y_t-aY_{t-1})/\sqrt{\omega+bY_{t-1}^2}\right|^{\alpha}-1,$$
$$X_{t,2}(\boldsymbol{\theta})=\big(\partial(Z_t(\omega,a,b))/\partial\omega\big)\mathrm{sgn}\{Z_t(\omega,a,b)\},$$
$$X_{t,3}(\boldsymbol{\theta})=\big(\partial(Z_t(\omega,a,b))/\partial a\big)\mathrm{sgn}\{Z_t(\omega,a,b)\},$$
$$X_{t,4}(\boldsymbol{\theta})=\big(\partial(Z_t(\omega,a,b))/\partial b\big)\mathrm{sgn}\{Z_t(\omega,a,b)\},$$

and write

$$\mathbf{X}_t(\boldsymbol{\theta}, \alpha) = \left(X_{t,1}(\boldsymbol{\theta}, \alpha), X_{t,2}(\boldsymbol{\theta}), X_{t,3}(\boldsymbol{\theta}), X_{t,4}(\boldsymbol{\theta})\right)^T.$$

Based on the estimating equations $\sum_{t=1}^{n} \mathbf{X}_t(\boldsymbol{\theta}, \alpha) = 0$, we define the empirical likelihood function of $\boldsymbol{\theta}$ and $\alpha$ as

$$L(\boldsymbol{\theta}, \alpha) = \sup\left\{\prod_{t=1}^{n}(np_t) : p_1 \geq 0, \ldots, p_n \geq 0, \sum_{t=1}^{n} p_t = 1, \sum_{t=1}^{n} p_t \mathbf{X}_t(\boldsymbol{\theta}, \alpha) = 0\right\}.$$

**Theorem 3.11.** *Under conditions of Theorem 3.10,* $-2\log\left(\max_{\boldsymbol{\theta}} L(\boldsymbol{\theta}, \alpha_0)\right)$ *converges in distribution to a chi-squared distribution with one degree of freedom as* $n \to \infty$.

*Proof.* See Zhang et al. [106]. □

## 3.7 CONDITIONAL VALUE-AT-RISK

Value-at-Risk (VaR) and expected shortfall are two commonly employed risk measures in insurance and finance. When observations follow from a time series model, corresponding conditional risk measures are useful. Here we focus on the conditional VaR for a GARCH sequence.

Suppose our observations $X_1, \cdots, X_n$ follow from a GARCH$(p, q)$ model:

$$X_t = \sigma_t^* \varepsilon_t^*, \quad (\sigma_t^*)^2 = \alpha_0^* + \sum_{i=1}^{p} \alpha_i^* X_{t-i}^2 + \sum_{j=1}^{q} \beta_j^* (\sigma_{t-j}^*)^2, \tag{3.36}$$

where $\{\varepsilon_t^*\}$ is a sequence of independent and identically distributed random variables with mean zero and variance one. Then, for $r \in (0, 1)$, the one-step ahead $100r\%$ conditional VaR, given $X_1, \cdots, X_n$, is defined as

$$q_{r,n} := \inf\{x : P(X_{n+1} \leq x | X_{n+1-k}, k \geq 1) \geq r\} = \sigma_{n+1}^* \theta_{\varepsilon,r}^*, \tag{3.37}$$

where $\theta_{\varepsilon,r}^*$ denotes the $r$-th quantile of $\varepsilon_t^*$. Therefore a simple nonparametric estimator for the conditional VaR $q_{r,n}$ is

$$\hat{q}_{r,n} = \hat{\sigma}_{n+1}^* \hat{\theta}_{\varepsilon,r}^*,$$

where $\hat{\sigma}_t^*$ is an estimator of the conditional standard deviation at time $t$ and $\hat{\theta}_{\varepsilon,r}^*$ is an estimator of the $100r\%$ quantile of $\varepsilon_t^*$. To get $\hat{\sigma}_t^*$, one can simply replace the parameters in the GARCH model by some corresponding estimators. To ensure that the process $\{X_t, t = 0, \pm 1, \pm 2, \ldots, \}$, defined by Eq. (3.36), is strictly stationary with $EX_t^2 < \infty$, we need the following assumption.

**F1)** For each $\boldsymbol{\gamma}^* = (\alpha_0^*, \alpha_1^*, \cdots, \alpha_p^*, \beta_1^*, \cdots, \beta_q^*)^T \in \Theta$, $\alpha_0^* > 0$, $\alpha_i^* > 0$ for $i = 1, \cdots, p$, $\beta_j^* \geq 0$ for $j = 1, \cdots, q$, and $\sum_{i=1}^p \alpha_i^* + \sum_{j=1}^q \beta_j^* < 1$.

Under the above assumption, $\sigma_t^*$ may be expressed as

$$(\sigma_t^*)^2 := (\sigma_t^*(\boldsymbol{\gamma}^*))^2 = \frac{\alpha_0^*}{1 - \sum_{j=1}^q \beta_j^*} + \sum_{i=1}^p \alpha_i^* X_{t-i}^2 + \sum_{i=1}^p \alpha_i^* \sum_{k=1}^\infty \sum_{j_1=1}^q \cdots \sum_{j_k=1}^q \beta_{j_1}^* \cdots \beta_{j_k}^* X_{t-i-j_1-\cdots-j_k}^2.$$

If one employs the quasi-maximum likelihood estimator for $\boldsymbol{\gamma}^*$ in estimating the conditional VaR, then $E(\varepsilon_t^{*4}) < \infty$ is needed to ensure a normal limit. Here we propose to employ the least absolute deviations estimator in Peng and Yao [85] instead of QMLE so as to allow $\varepsilon_t^*$ to have a heavier tail.

Assume the unknown median of $(\varepsilon_t^*)^2$ is $d > 0$ and put $\varepsilon_t = \varepsilon_t^*/\sqrt{d}$. Then the median of $\log \varepsilon_t^2$ becomes $\log\{\text{median}((\varepsilon_t^*)^2/d)\} = 0$. Furthermore, model (3.36) can be written as

$$X_t = \sigma_t \varepsilon_t, \quad \sigma_t^2 = \alpha_0 + \sum_{i=1}^p \alpha_i X_{t-i}^2 + \sum_{j=1}^q \beta_j \sigma_{t-j}^2, \tag{3.38}$$

where $\sigma_t = \sqrt{d}\sigma_t^*$, $\alpha_0 = d\alpha_0^*$, $\alpha_i = d\alpha_i^*$ for $i = 1, \cdots, p$, and $\beta_j = \beta_j^*$ for $j = 1, \cdots, q$.

Put $\boldsymbol{\gamma} = (\alpha_0, \alpha_1, \cdots, \alpha_p, \beta_1, \cdots, \beta_q)^T$. Then

$$\sigma_t^2 := \sigma_t^2(\boldsymbol{\gamma}) = \frac{\alpha_0}{1 - \sum_{j=1}^q \beta_j} + \sum_{i=1}^p \alpha_i X_{t-i}^2 + \sum_{i=1}^p \alpha_i \sum_{k=1}^\infty \sum_{j_1=1}^q \cdots \sum_{j_k=1}^q \beta_{j_1} \cdots \beta_{j_k} X_{t-i-j_1-\cdots-j_k}^2.$$

Since the above formula involves $X_j$ for $j < 0$, we use the following truncated version in inference:

$$\bar{\sigma}_t^2(\boldsymbol{\gamma}) = \frac{\alpha_0}{1-\sum_{j=1}^q \beta_j} + \sum_{i=1}^{\min(p,t-1)} \alpha_i X_{t-i}^2 + \sum_{i=1}^p \alpha_i \sum_{k=1}^\infty \sum_{j_1=1}^q \cdots \sum_{j_k=1}^q \beta_{j_1}\cdots\beta_{j_k} X_{t-i-j_1-\cdots-j_k}^2 I(t-i-j_1-\cdots-j_k \ge 1).$$

Therefore the least absolute deviations estimator for $\boldsymbol{\gamma}$ is defined as

$$\hat{\boldsymbol{\gamma}} = \arg\min_{\boldsymbol{\gamma}} \sum_{t=1}^n \left|\log X_t^2 - \log \bar{\sigma}_t^2(\boldsymbol{\gamma})\right|.$$

Using $\hat{\boldsymbol{\gamma}}$, we estimate $\varepsilon_t$ by $\hat{\varepsilon}_t = X_t/\bar{\sigma}_t(\hat{\boldsymbol{\gamma}})$ and let $\hat{\theta}_{\varepsilon,r}$ denote the $r$-th sample quantile of $\hat{\varepsilon}_1, \cdots, \hat{\varepsilon}_n$. Finally $q_{r,n}$ is estimated by

$$\hat{q}_{r,n} = \bar{\sigma}_{n+1}(\hat{\boldsymbol{\gamma}})\hat{\theta}_{\varepsilon,r}.$$

In order to derive the asymptotic distribution of the above proposed conditional VaR estimator, we consider the conditional VaR at a fixed level and an extreme level separately.
**Case 1: Fixed level.** First we consider the case of a fixed level, i.e., $r \in (0,1)$ is a constant. In this case $\hat{\theta}_{\varepsilon,r} = \hat{\varepsilon}_{n,[nr]}$, where $\hat{\varepsilon}_{n,1} \le \cdots \le \hat{\varepsilon}_{n,n}$ denote the order statistics of $\hat{\varepsilon}_1, \cdots, \hat{\varepsilon}_n$.

**Theorem 3.12.** *Suppose model (3.36) holds with F1) above, $r \in (0,1)$ is a fixed level and the density function of $\log \varepsilon_t^2$ is positive and differentiable at zero. Then, as $n \to \infty$, $\sqrt{n}\{\hat{q}_{r,n} - q_{r,n}\}$ converges in distribution to a normal limit with mean zero and a complicated variance given in the proof below.*

*Proof.* Define $A_t(\boldsymbol{\gamma}) = (A_{t,1}(\boldsymbol{\gamma}), \cdots, A_{t,p+q+1}(\boldsymbol{\gamma}))^T$, where

$$A_{t,1}(\boldsymbol{\gamma}) = \frac{1}{\sigma_t^2(\boldsymbol{\gamma})}\frac{1}{1-\sum_{j=1}^q \beta_j},$$

$$A_{t,1+i}(\boldsymbol{\gamma}) = \frac{1}{\sigma_t^2(\boldsymbol{\gamma})}\{X_{t-i}^2 + \sum_{k=1}^\infty \sum_{j_1=1}^q \cdots \sum_{j_k=1}^q \beta_{j_1}\cdots\beta_{j_k} X_{t-i-j_1-\cdots-j_k}^2\} \quad \text{for} \quad i = 1, \cdots, p,$$

$$A_{t,p+1+j}(\boldsymbol{\gamma}) = \frac{1}{\sigma_t^2(\boldsymbol{\gamma})}\{\frac{\alpha_0}{(1-\sum_{j=1}^q \beta_j)^2} + \sum_{i=1}^p \alpha_i X_{t-i-j}^2 + \sum_{i=1}^p \alpha_i \sum_{k=1}^\infty (k+1) \sum_{j_1=1}^q \cdots \sum_{j_k=1}^q \beta_{j_1}\cdots\beta_{j_k} X_{t-i-j-j_1-\cdots-j_k}^2\}$$

for $j = 1, \cdots, q$. Then it follows from Peng and Yao [85] that

$$\sqrt{n}\left\{\hat{\boldsymbol{\gamma}} - \boldsymbol{\gamma}_0\right\} = \frac{\Sigma^{-1}}{2g(0)} \frac{1}{\sqrt{n}} \sum_{t=1}^{n} A_t(\boldsymbol{\gamma}_0) sgn(\log \varepsilon_t^2) + o_p(1), \tag{3.39}$$

where $\Sigma = E\{A_t(\boldsymbol{\gamma}_0)A_t^T(\boldsymbol{\gamma}_0)\}$ and $g(x)$ denotes the density function of $\log \varepsilon_t^2$.

Write

$$\begin{aligned}
&\frac{1}{n}\sum_{t=1}^{n} I(X_t/\hat{\sigma}_t \le x) - F_\varepsilon(x) \\
&= \frac{1}{n}\sum_{t=1}^{n}\left\{I(X_t/\sigma_t \le x\hat{\sigma}_t/\sigma_t) - I(X_t/\sigma_t \le x) - F_\varepsilon(x\hat{\sigma}_t/\sigma_t) + F_\varepsilon(x)\right\} \\
&\quad + \frac{1}{n}\sum_{t=1}^{n}\{I(X_t/\sigma_t \le x) - F_\varepsilon(x)\} + \frac{1}{n}\sum_{t=1}^{n}\{F_\varepsilon(x\hat{\sigma}_t/\sigma_t) - F_\varepsilon(x)\}.
\end{aligned}$$

Like the proof of Berkes and Horváth [7], we can show that

$$\begin{aligned}
&\frac{1}{\sqrt{n}}\sum_{t=1}^{n}\{I(X_t/\hat{\sigma}_t \le x) - F_\varepsilon(x)\} \\
&= \frac{1}{\sqrt{n}}\sum_{t=1}^{n}\{I(\varepsilon_t \le x) - F_\varepsilon(x)\} + \sqrt{n}(\hat{\boldsymbol{\gamma}} - \boldsymbol{\gamma}_0)^T E(\frac{\partial \sigma_t/\partial \boldsymbol{\gamma}}{\sigma_t}) F_\varepsilon'(x)x + o_p(1)
\end{aligned}$$

uniformly in a neighborhood of $F_\varepsilon^-(r)$. That is

$$\begin{aligned}
&\sqrt{n}(\frac{[nr]}{n} - F_\varepsilon(\hat{\theta}_{\varepsilon,r})) \\
&= \frac{1}{\sqrt{n}}\sum_{t=1}^{n}\{I(\varepsilon_t \le F_\varepsilon^-(r)) - r\} + \sqrt{n}(\hat{\boldsymbol{\gamma}} - \boldsymbol{\gamma}_0)^T E(\frac{\partial \sigma_t/\partial \boldsymbol{\gamma}}{\sigma_t}) F_\varepsilon'(F_\varepsilon^-(r)) F_\varepsilon^-(r) \\
&\quad + o_p(1),
\end{aligned}$$

which implies that

$$\begin{aligned}
&\sqrt{n}(\hat{\theta}_{\varepsilon,r} - F_\varepsilon^-(r)) \\
&= -\frac{1}{F_\varepsilon'(F_\varepsilon^-(r))}\frac{1}{\sqrt{n}}\sum_{t=1}^{n}\{I(\varepsilon_t \le F_\varepsilon^-(r)) - r\} - \sqrt{n}(\hat{\boldsymbol{\gamma}} - \boldsymbol{\gamma}_0)^T E(\frac{\partial \sigma_t/\partial \boldsymbol{\gamma}}{\sigma_t}) F_\varepsilon^-(r) \\
&\quad + o_p(1).
\end{aligned}$$

Therefore

$$\sqrt{n}\{\hat{q}_{r,n} - q_{r,n}\} = \sqrt{n}(\hat{\boldsymbol{\gamma}} - \boldsymbol{\gamma}_0)^T \frac{\partial \sigma_{n+1}(\boldsymbol{\gamma}_0)}{\partial \boldsymbol{\gamma}} F_\varepsilon^-(r) + \sigma_{n+1}(\boldsymbol{\gamma}_0)\sqrt{n}\{\hat{\theta}_{\varepsilon,r} - F_\varepsilon^-(r)\}$$

$$= \{\frac{\partial \sigma_{n+1}(\boldsymbol{\gamma}_0)}{\partial \boldsymbol{\gamma}} - \sigma_{n+1}(\boldsymbol{\gamma}_0) E(\frac{\partial \sigma_1(\boldsymbol{\gamma}_0)/\partial \boldsymbol{\gamma}}{\sigma_1(\boldsymbol{\gamma}_0)})\}^T$$

$$\times F_\varepsilon^-(r) \frac{\Sigma^{-1}}{2g(0)} \frac{1}{\sqrt{n}} \sum_{t=1}^n A_t(\boldsymbol{\gamma}_0) sgn(\log \epsilon_t^2)$$

$$- \frac{\sigma_{n+1}(\boldsymbol{\gamma}_0)}{F_\varepsilon'(F_\varepsilon^-(r))} \frac{1}{\sqrt{n}} \sum_{t=1}^n \{I(\varepsilon_t \le F_\varepsilon^-(r)) - r\} + o_p(1).$$

Hence the theorem follows from the above expansion, where the asymptotic variance can be computed and stated in a complicated formula. □

Since the above asymptotic variance is too complicated, constructing an interval via estimating the asymptotic variance becomes nontrivial. Here we develop an empirical likelihood method to construct an interval for $q_{r,n}$.

First note that $\hat{\boldsymbol{\gamma}}$ is equivalent to solving the following score equations

$$\sum_{t=1}^n D_t(\boldsymbol{\gamma}) = 0 \quad \text{with} \quad D_t(\boldsymbol{\gamma}) = \frac{\partial \log \bar{\sigma}_t^2(\boldsymbol{\gamma})}{\partial \boldsymbol{\gamma}} sgn(\log \frac{X_t^2}{\bar{\sigma}^2(\boldsymbol{\gamma})}).$$

Let $K$ be a symmetric density function with support in $[-1, 1]$ and have a continuous first derivative. Put $G(x) = \int_{-1}^x K(y)dy$ for $x \in [-1, 1]$ and define

$$g_t(\theta, \boldsymbol{\gamma}) = \left(G(\frac{\theta/\sqrt{\bar{\sigma}_{n+1}(\boldsymbol{\gamma})} - \varepsilon_t(\boldsymbol{\gamma})}{h}) - r, D_t^T(\boldsymbol{\gamma})\right)^T = \left(\omega_t(\boldsymbol{\gamma}), D_t^T(\boldsymbol{\gamma})\right)^T. \tag{3.40}$$

Then the empirical likelihood function for $\theta$ and $\boldsymbol{\gamma}$ is defined as

$$L_n(\theta, \boldsymbol{\gamma}) = \sup \left\{ \prod_{t=1}^n (np_t) : p_1 \ge 0, \cdots, p_n \ge 0, \sum_{t=1}^n p_t = 1, \sum_{t=1}^n p_t g_t(\theta, \boldsymbol{\gamma}) = 0 \right\}.$$

It follows from the Lagrange multiplier technique that

$$p_t = \frac{1}{n\left\{1 + \boldsymbol{\lambda}^T g_t(\theta, \boldsymbol{\gamma})\right\}} \quad \text{for} \quad t = 1, \cdots, n,$$

and

$$l_n(\theta, \boldsymbol{\gamma}) = -2 \log L_n(\theta, \boldsymbol{\gamma}) = 2 \sum_{t=1}^n \log\left\{1 + \boldsymbol{\lambda}^T g_t(\theta, \boldsymbol{\gamma})\right\},$$

where $\boldsymbol{\lambda} = \boldsymbol{\lambda}(\theta, \boldsymbol{\gamma})$ satisfies

$$\sum_{t=1}^{n} \frac{g_t(\theta, \boldsymbol{\gamma})}{1 + \boldsymbol{\lambda}^T g_t(\theta, \boldsymbol{\gamma})} = 0.$$

Since we are interested in constructing a confidence interval for $\theta$, we consider the profile empirical likelihood function

$$l_n^P(\theta) = \min_{\gamma} l_n(\theta, \boldsymbol{\gamma}).$$

Throughout we use $\theta_0$ to denote the true value of $\theta = q_{r,n}$ and write $\sigma_t = \sigma_t(\boldsymbol{\gamma})$. Using the standard arguments in Qin and Lawless [89], we can show the following theorem.

**Theorem 3.13.** *Suppose condition F1) holds and* $E(|\varepsilon_t|^{2+\delta}) < \infty$ *for some* $\delta > 0$. *Further assume the density function of* $\varepsilon_t$ *is positive and differentiable at* $\theta_0/\sigma_{n+1}(\boldsymbol{\gamma}_0)$, *the density function of* $\log \varepsilon_t^2$ *is positive and differentiable at zero,* $n^{1-d}h^2 \to \infty$ *and* $nh^4 \to 0$ *for some* $d \in (0, 1/2)$ *as* $n \to \infty$. *Then,*

$$l_n^P(\theta_0) \xrightarrow{d} \chi_1^2 \quad as \quad n \to \infty.$$

*Proof.* It can be proved by using Theorem 3.12 and the standard arguments in Qin and Lawless [89]. See Gong et al. [45] for details. □

Based on the above theorem, a confidence interval of $\theta_0$ with level $\xi$ can be obtained as

$$I_\xi(r) = \{\theta : l_n^P(\theta) \le \chi_{1,\xi}^2\},$$

where $\chi_{1,\xi}^2$ is the $\xi$th quantile of a chi-squared distribution with one degree of freedom.

Motivated by the optimal choice of bandwidth in smoothing distribution estimation (see Cheng and Peng [21]), the bandwidth is chosen as $h = cn^{-1/3}$ for some positive $c$ in practice.

**Case 2: Extreme level.** Here we consider an extreme level, i.e.,

$$r = r(n) \to 1, \quad n(1 - r) \to d \in [0, \infty) \quad \text{as} \quad n \to \infty.$$

In this case, one has to extrapolate the distribution function of $\varepsilon_t$. When $\varepsilon_t$ has a heavy tail satisfying (2.1), the conditional VaR $q_{r,n}$ in (3.37) can be estimated by

$$\tilde{q}_{r,n} = \bar{\sigma}_{n+1}(\hat{\boldsymbol{\gamma}})\hat{\varepsilon}_{n,n-k}(\frac{n(1-r)}{k})^{-1/\hat{\alpha}_\varepsilon},$$

where

$$\hat{\alpha}_\varepsilon = \left\{ \frac{1}{k} \sum_{i=1}^{k} \log \frac{\hat{\varepsilon}_{n,n-i+1}}{\hat{\varepsilon}_{n,n-k}} \right\}^{-1},$$

and $\hat{\varepsilon}_{n,1} \leq \cdots \leq \hat{\varepsilon}_{n,n}$ denote the order statistics of $\hat{\varepsilon}_1, \cdots, \hat{\varepsilon}_n$.

**Theorem 3.14.** *Suppose model (3.36) holds with F1) above, and $r \in (0, 1)$ is an extreme level satisfying*

$$1 - r \to 0, \quad \frac{n(1-r)}{k} \to 0, \quad \log\big(n(1-r)\big) = o(\sqrt{k}) \quad as \quad n \to \infty.$$

*Assume conditions (2.4) for the distribution function of $\varepsilon_t$, (2.28), and*

$$\sqrt{k}\frac{A(k/n)}{(k/n)^{1/\alpha}\bar{F}^{-}(k/n)} \to \lambda \in \mathbb{R} \quad as \quad n \to \infty.$$

*Further assume the density function of $\log \varepsilon_t^2$ is positive and differentiable at zero. Then*

$$\frac{\sqrt{k}}{\log \frac{n(1-r)}{k}} \left\{ \frac{\tilde{q}_{r,n}}{q_{r,n}} - 1 \right\} \xrightarrow{d} N(\frac{\lambda}{1+\rho}, 1/\alpha^2) \quad as \quad n \to \infty.$$

*Proof.* It follows from Theorem 2.16 and (3.39). □

**Remark 3.2.** Like Chan et al. [16], an empirical likelihood method can be developed to construct a confidence interval for the above conditional Value-at-Risk at an extreme level.

## 3.8 HEAVY TAILED AR–GARCH SEQUENCES

Consider the following strictly stationary AR($s$)–GARCH(1, 1) process

$$\begin{cases} Y_t = \sum_{i=1}^{s} \phi_i Y_{t-i} + \varepsilon_t, \\ \varepsilon_t = h_t \eta_t, \quad h_t^2 = \omega + \alpha \varepsilon_{t-1}^2 + \beta h_{t-1}^2, \end{cases} \tag{3.41}$$

where $\omega > 0$, $\alpha \geq 0$, $\beta \geq 0$, $\alpha + \beta > 0$, $\{\eta_t\}$ is a sequence of independent and identically distributed random variables with mean zero and variance one.

Recently, Francq and Zakoïan [39] showed that the quasi-maximum likelihood estimator for jointly estimating all parameters has a normal limit when both $EY_t^4 < \infty$ and $E\varepsilon_t^4 < \infty$. A computationally efficient estimation procedure is to first estimate $\boldsymbol{\phi} = (\phi_1, \cdots, \phi_s)^T$ by the least squares estimator, and then estimate the parameters in the GARCH part by the quasi-maximum likelihood estimator. That is to estimate $\boldsymbol{\phi}$ by

$$\hat{\boldsymbol{\phi}} = \left\{ \frac{1}{n} \sum_{t=1}^{n} \boldsymbol{Z}_t \boldsymbol{Z}_t^T \right\}^{-1} \frac{1}{n} \sum_{t=1}^{n} Y_t \boldsymbol{Z}_t, \quad \text{where} \quad \boldsymbol{Z}_t = (Y_{t-1}, \cdots, Y_{t-s})^T.$$

It is known that $\sqrt{n}\{\hat{\boldsymbol{\phi}} - \boldsymbol{\phi}\}$ has a normal limit when $\{\varepsilon_t\}$ is a sequence of independent and identically distributed random variables with mean zero and finite variance. Hence, one may wonder what the limit of the least squares estimator $\hat{\boldsymbol{\phi}}$ for $\boldsymbol{\phi}$ is under model (3.41). Recently Lange [66] showed that the limit is nonnormal when $E\varepsilon_t^2 < \infty$, but $E\varepsilon_t^4 = \infty$ under the following regularity conditions.

**G1)** $\eta_t$ has a density function with respect to the Lebesgue measure on $\mathbb{R}$ that is bounded away from zero and infinity on compact sets.

**G2)** $E\{\log(\beta + \alpha\eta_t^2)\} < 0$.

**G3)** The maximal eigenvalue of the companion form matrix corresponding to the AR part of the model is smaller than one.

**G4)** The initial values are distributed according to the stationary distribution.

**G5)** The GARCH$(1, 1)$ process has a finite second order moment, but infinite fourth order moment.

**G6)** $\eta_t$ has a symmetric density function.

**Theorem 3.15.** *Assume model (3.41) holds with conditions G1)–G6) above. Then $(Y_t, \cdots, Y_{t-k})^T$ is a regular variation with index $\gamma$ for $k = 0$ and a multivariate regular variation (defined in the next chapter) with index $\gamma$ for any $k \geq 1$, and*

$$n^{1-\gamma/2}\{\hat{\boldsymbol{\phi}} - \boldsymbol{\phi}\} \xrightarrow{d} \{E(\boldsymbol{Z}_1 \boldsymbol{Z}_1^T)\}^{-1} \boldsymbol{S} \quad \text{as} \quad n \to \infty,$$

*where $\boldsymbol{S}$ is a stable random vector with index $\gamma/2$ and $\gamma$ satisfies $E\{(\beta + \alpha\eta_t^2)^{\gamma/2}\} = 1$.*

*Proof.* See the proofs of Theorems 1 and 2 in Lange [66]. □

To investigate the limit of $\hat{\boldsymbol{\phi}}$ for the case of $E\varepsilon_t^2 = \infty$, Zhang and Ling [105] consider the following AR$(s)$ model with G–GARCH$(1, 1)$ errors:

$$\begin{cases} Y_t = \sum_{i=1}^{s} \phi_i Y_{t-i} + \varepsilon_t, \\ \varepsilon_t = h_t \eta_t, \quad h_t^\delta = g(\eta_{t-1}) + c(\eta_{t-1}) h_{t-1}^\delta, \end{cases} \tag{3.42}$$

where $\delta > 0$, $P(h_t^\delta > 0) = 1$, $c(0) < 1$, $c(\cdot)$ and $g(\cdot)$ are nonnegative functions, and $\{\eta_t\}$ is a sequence of independent and identically distributed with symmetric distribution.

Assume the following regularity conditions.

**H1)** $E\{\log(c(\eta_t))\} < 0$.

**H2)** There exists a $\kappa > 0$ such that

$$E(c(\eta_t))^\kappa \geq 1, \quad E\{(c(\eta_t))^\kappa \log^+(c(\eta_t))\} < \infty,$$
$$\text{and} \quad E(g(\eta_t) + |\eta_t|^\delta)^\kappa < \infty,$$

where $\log^+(x) = \log(\max(e, x))$.

**H3)** The density of $\eta_t$ is positive in the neighborhood of zero.

First Zhang and Ling [105] proved that there exists a unique $\gamma \in (0, \delta\kappa)$ such that

$$E(c(\eta_t))^{\gamma/\delta} = 1 \quad \text{and} \quad P(|\varepsilon_t| > x) \sim c_0^{(\gamma)} x^{-\gamma} E(|\eta_t|^\gamma) \quad \text{as} \quad x \to \infty,$$

where

$$c_0^{(\gamma)} = \frac{E\{(g(\eta_1) + c(\eta_1)\sigma_1^\delta)^{\gamma/\delta} - (c(\eta_1)\sigma_1^\delta)^{\gamma/\delta}\}}{\gamma E\{(c(\eta_1))^{\gamma/\delta} \log^+(c(\eta_1))\}}.$$

Further Zhang and Ling [105] derived the asymptotic distribution of $\hat{\boldsymbol{\phi}}$ for $\gamma \in (0, 4]$ as follows.

**Theorem 3.16.** *Assume model (3.42) holds with the above conditions H1)–H3) and G3). Then*

**i)** *when $\gamma \in (0, 2)$, $\hat{\boldsymbol{\phi}} - \boldsymbol{\phi} \xrightarrow{d} \Sigma_{\gamma/2}^{-1} \mathbf{S}_{\gamma/2}$, where $\mathbf{S}_{\gamma/2}$ is a s-dimensional stable vector with index $\gamma/2$ and $\Sigma_{\gamma/2}$ is a $s \times s$ matrix whose elements are composed of stable variables with index $\gamma/2$;*

**ii)** *when $\gamma = 2$, $\log(n)\{\hat{\boldsymbol{\phi}} - \boldsymbol{\phi}\} \xrightarrow{d} \left(\sum_{l=0}^\infty \psi_l \psi_{l+|i-j|}\right)_{s\times s}^{-1} \mathbf{S}_{\gamma/2}$, where $\psi_l$ is defined in $Y_t = \sum_{l=0}^\infty \psi_l \varepsilon_{t-l}$;*

**iii)** *when $2 < \gamma < 4$, $n^{1-2\gamma}\{\hat{\boldsymbol{\phi}} - \boldsymbol{\phi}\} \xrightarrow{d} \Sigma^{-1} \mathbf{S}_{\gamma/2}$, where $\Sigma = E(\mathbf{Z}_1 \mathbf{Z}_1^T)$;*

**iv)** *when $\gamma = 4$, $\sqrt{n/\log(n)}\{\hat{\boldsymbol{\phi}} - \boldsymbol{\phi}\} \xrightarrow{d} \Sigma^{-1} N(0, A)$, where $A = (c_0^{(4)} E(\eta_1^2))(a_{ij})_{s\times s}$ is positive definite with*

$$\begin{cases} a_{ij} = \lim_{M\to\infty} E\{u_{t,i,M} u_{t,j,M}\}, \\ u_{t,i,M} = \sum_{l=i}^M \psi_{l-i} \eta_{t-l} \prod_{j=1}^l c^{1/\delta}(\eta_{t-j}) \prod_{k=l+1}^M c^{2/\delta}(\eta_{t-k}). \end{cases}$$

*Proof.* See Theorem 2.1 of Zhang and Ling [105]. □

The above theorem shows that the least squares estimator is inconsistent when $\gamma \in (0,2)$, and has a different limit for the cases of $\gamma = 2$, $\gamma \in (2,4)$, $\gamma = 4$ and $\gamma > 4$. Therefore interval estimation for $\boldsymbol{\phi}$ becomes quite challenging. To avoid the nonnormal limit, Zhu and Ling [108] proposed the following weighted least absolute deviations estimator

$$\tilde{\boldsymbol{\phi}} = \arg\min_{\boldsymbol{\phi}} \frac{1}{n}\sum_{t=1}^{n} w_t |Y_t - \sum_{i=1}^{s} \phi_i Y_{t-i}|,$$

where

$$w_t = \left\{ \max\left(1,\, C^{-1}\sum_{k=1}^{p} k^{-9}|Y_{t-k}|I(|Y_{t-k}| > C)\right)\right\}^{-1}$$

and $C > 0$ is taken as 90% or 95% sample quantile of $Y_1, \cdots, Y_n$.

**Theorem 3.17.** *Consider*

$$Y_t = \sum_{i=1}^{s} \phi_i Y_{t-i} + \varepsilon_t, \qquad \varepsilon_t = h_t \eta_t.$$

*Assume G3), $\varepsilon_t$ is strictly stationary and ergodic, $E|\varepsilon_t|^{2\tau_0} < \infty$ for some $\tau_0 \in (0,1)$, $h_t \geq c_0$ almost surely for some positive constant $c_0$, $\eta_t$ has a zero median with a continuous density function $g(x)$ satisfying $g(0) > 0$ and $\sup_{x\in\mathbb{R}} g(x) < \infty$. Then*

$$\sqrt{n}\left\{\tilde{\boldsymbol{\phi}} - \boldsymbol{\phi}\right\} \xrightarrow{d} N\left(0, (2g(0))^2 \Sigma^{-1}\Omega\Sigma^{-1}\right) \quad \text{as} \quad n \to \infty,$$

*where*

$$\Omega = E\left\{w_t^2 \frac{\partial \varepsilon_t}{\partial \boldsymbol{\phi}}\frac{\partial \varepsilon_t}{\partial \boldsymbol{\phi}^T}\right\} \quad \text{and} \quad \Sigma = E\left\{\frac{w_t}{h_t}\frac{\partial \varepsilon_t}{\partial \boldsymbol{\phi}}\frac{\partial \varepsilon_t}{\partial \boldsymbol{\phi}^T}\right\}.$$

*Proof.* See Theorem 2 of Zhu and Ling [108]. □

## 3.9 SELF-WEIGHTED ESTIMATION FOR ARMA–GARCH MODELS

Assume that $\{Y_t\}_{t=-\infty}^{\infty}$ are generated by the following ARMA–GARCH model:

$$Y_t = \mu + \sum_{i=1}^{p} \phi_i Y_{t-i} + \sum_{j=1}^{q} \psi_j \varepsilon_{t-j} + \varepsilon_t, \tag{3.43}$$

$$\varepsilon_t = \eta_t\sqrt{h_t} \quad \text{and} \quad h_t = \alpha_0 + \sum_{i=1}^{r}\alpha_i\varepsilon_{t-i}^2 + \sum_{j=1}^{s}\beta_j h_{t-j}, \tag{3.44}$$

where $\alpha_0 > 0, \alpha_i \geq 0$ for $i = 1, \cdots, r$, $\beta_j \geq 0$ for $j = 1, \cdots, s$, and $\{\eta_t\}$ is a sequence of i.i.d. random variables with mean zero and variance one. Denote

$$\boldsymbol{\gamma} = (\mu, \phi_1, \cdots, \phi_p, \psi_1, \cdots, \psi_q)^T, \quad \boldsymbol{\delta} = (\alpha_0, \alpha_1, \cdots, \alpha_r, \beta_1, \cdots, \beta_s)^T,$$
$$\boldsymbol{\theta} = (\boldsymbol{\gamma}^T, \boldsymbol{\delta}^T)^T.$$

The parameter subspaces, $\Theta_{\boldsymbol{\gamma}} \subset \mathbb{R}^{p+q+1}$ and $\Theta_{\boldsymbol{\delta}} \subset \mathbb{R}_+^{r+s+1}$, are compact sets, where $\mathbb{R}_+ = [0, \infty)$. Let $\Theta = \Theta_{\boldsymbol{\gamma}} \times \Theta_{\boldsymbol{\delta}}$, $m = p + q + s + r + 2$ and $\boldsymbol{\theta}_0$ be the true value of $\boldsymbol{\theta}$. Denote $\alpha(z) = \sum_{i=1}^{r}\alpha_i z^i$, $\beta(z) = 1 - \sum_{i=1}^{s}\beta_i z^i$, $\phi(z) = 1 - \sum_{i=1}^{p}\phi_i z^i$ and $\psi(z) = 1 + \sum_{i=1}^{q}\psi_i z^i$.

Given the observations $\{Y_1, \cdots, Y_n\}$ and the initial values $\{Y_0, Y_{-1}, \cdots\}$ which are generated by models (3.43) and (3.44), we can write the corresponding parametric model as

$$\varepsilon(\boldsymbol{\gamma}) = Y_t - \mu - \sum_{i=1}^{p}\phi_i Y_{t-i} - \sum_{j=1}^{q}\psi_j\varepsilon_{t-j}(\boldsymbol{\gamma}),$$

$$\eta_t(\boldsymbol{\theta}) = \varepsilon_t(\boldsymbol{\gamma})/\sqrt{h_t(\boldsymbol{\theta})}, \quad h_t(\boldsymbol{\theta}) = \alpha_0 + \sum_{i=1}^{r}\varepsilon_{t-i}^2(\boldsymbol{\gamma}) + \sum_{j=1}^{s}\beta_j h_{t-j}(\boldsymbol{\theta}).$$

Define $w_t = w(Y_{t-1}, Y_{t-2}, \cdots)$, where $w$ is a measurable function satisfying some conditions. Since we do not have the initial values $\{Y_i : i \leq 0\}$, we denote $\varepsilon_t(\boldsymbol{\gamma}), h_t(\boldsymbol{\theta}), w_t$ as $\tilde{\varepsilon}_t(\boldsymbol{\gamma}), \tilde{h}_t(\boldsymbol{\theta}), \tilde{w}_t$, respectively, when $\{Y_i : i \leq 0\}$ are replaced by a constant not depending on parameters (for example, zero). Therefore the self-weighted quasi-maximum likelihood estimator is defined as

$$\hat{\boldsymbol{\theta}}_{swqmle} = \arg\min_{\boldsymbol{\theta}} \sum_{t=1}^{n}\tilde{w}_t\left\{\log\tilde{h}_t(\boldsymbol{\theta}) + \frac{\tilde{\varepsilon}_t^2(\boldsymbol{\gamma})}{\tilde{h}_t(\boldsymbol{\theta})}\right\}.$$

To derive the asymptotic distribution of the above estimator, assume the following conditions.

**I1)** $\boldsymbol{\theta}_0$ is an interior point in $\boldsymbol{\Theta}$ and for each $\boldsymbol{\theta} \in \Theta$, $\phi(z) \neq 0$ and $\psi(z) \neq 0$ for $|z| \leq 1$, and $\phi(z)$ and $\psi(z)$ have no common root with $\phi_p \neq 0$ or $\psi_q \neq 0$.

**I2)** $\alpha(z)$ and $\beta(z)$ have no common root, $\alpha(1) \neq 0$, $\alpha_r + \beta_s \neq 0$ and $\sum_{i=1}^{s}\beta_i < 1$ for each $\boldsymbol{\theta} \in \Theta$.

**I3)** $\eta_t^2$ has a nondegenerate distribution with $E(\eta_t^2) = 1$.
**I4)** $E(|\varepsilon_t|^{2\tau}) < \infty$ for some $\tau > 0$.
**I5)** $w$ is a measurable, positive and bounded function satisfying

$$E\big(w_t\{1 + \sum_{i=0}^{\infty} \rho^i |Y_{t-i}|\}^4\big) < \infty \quad \text{for any} \quad \rho \in (0, 1).$$

**I6)** $E\big(|w_t - \tilde{w}_t|^{\tau_0/4}\big) = O(t^{-2})$ as $t \to \infty$, where $\tau_0 = \min(\tau, 1)$.
**I7)** $E(\eta_t^4) < \infty$ and $P(\eta_t^2 - c\eta_t - 1 = 0) < 1$ for any $c \in \mathbb{R}$.

**Theorem 3.18.** *Under conditions I1)–I7), we have*

$$\sqrt{n}\left\{\hat{\boldsymbol{\theta}}_{swqmle} - \boldsymbol{\theta}_0\right\} \xrightarrow{d} N\big(0, \Sigma_0^{-1}\Omega_0\Sigma_0^{-1}\big) \quad as \quad n \to \infty,$$

*where*

$$\Sigma_0 = E\big(w_t U_t(\boldsymbol{\theta}_0) U_t^T(\boldsymbol{\theta}_0)\big), \qquad \Omega_0 = E\big(w_t^2 U_t(\boldsymbol{\theta}_0) J U_t^T(\boldsymbol{\theta}_0)\big),$$

$$J = \begin{pmatrix} 1 & E(\eta_t^3)/\sqrt{2} \\ E(\eta_t^3)/\sqrt{2} & \big(E(\eta_t^4) - 1\big)/2 \end{pmatrix}, \quad U_t(\boldsymbol{\theta}) = \Big(h_t^{-1/2}\frac{\partial \varepsilon_t(\boldsymbol{\gamma})}{\partial \boldsymbol{\theta}}, (\sqrt{2}h_t)^{-1}\frac{\partial h_t(\boldsymbol{\theta})}{\partial \boldsymbol{\theta}}\Big).$$

*Proof.* See Theorem 3.1 of Ling [69]. □

Alternatively, when $\eta_t$ has median zero and $E(|\eta_t|) = 1$ instead of variance one, a so-called self-weighted quasi-maximum exponential likelihood estimator is defined as

$$\hat{\boldsymbol{\theta}}_{swqmele} = \arg\min_{\boldsymbol{\theta}} \sum_{t=1}^{n} \tilde{w}_t \left\{ \log\sqrt{\tilde{h}_t(\boldsymbol{\theta})} + \frac{|\varepsilon_t(\boldsymbol{\gamma})|}{\sqrt{\tilde{h}_t(\boldsymbol{\theta})}} \right\},$$

where $\tilde{w}_t$ and $\tilde{h}_t(\boldsymbol{\theta})$ are defined as those in $\hat{\boldsymbol{\theta}}_{swqmle}$.

To derive the asymptotic distribution of the above estimator, assume the following regularity conditions.

**J1)** $\boldsymbol{\theta}_0$ is an interior point in $\boldsymbol{\Theta}$ and for each $\boldsymbol{\theta} \in \Theta$, $\phi(z) \neq 0$ and $\psi(z) \neq 0$ for $|z| \leq 1$, and $\phi(z)$ and $\psi(z)$ have no common root with $\phi_p \neq 0$ or $\psi_q \neq 0$.
**J2)** $\alpha(z)$ and $\beta(z)$ have no common root, $\alpha(1) \neq 0$, $\alpha_r + \beta_s \neq 0$ and $\sum_{i=1}^{s} \beta_i < 1$ for each $\boldsymbol{\theta} \in \Theta$.
**J3)** $\eta_t^2$ has a nondegenerate distribution with $E(\eta_t^2) < \infty$.
**J4)** $E(|\varepsilon_t|^{2\tau}) < \infty$ for some $\tau > 0$.

**J5)** $w$ is a measurable, positive and bounded function satisfying

$$E\Big((w_t + w_t^2)\Big\{1 + \sum_{i=0}^{\infty} \rho^i |Y_{t-i}|\Big\}^3\Big) < \infty \quad \text{for any} \quad \rho \in (0,1).$$

**J6)** $E\big(|w_t - \tilde{w}_t|^{\tau_0/4}\big) = O(t^{-2})$ as $t \to \infty$, where $\tau_0 = \min(\tau, 1)$.

**J7)** $\eta_t$ has zero median with $E(|\eta_t|) = 1$ and a continuous density function $g(x)$ satisfying $g(0) > 0$ and $\sup_{x \in \mathbb{R}} g(x) < \infty$.

**Theorem 3.19.** *Under conditions J1)–J7), we have*

$$\sqrt{n}\Big\{\hat{\boldsymbol{\theta}}_{swqmele} - \boldsymbol{\theta}_0\Big\} \xrightarrow{d} N\big(0, \frac{1}{4}\bar{\Sigma}_0^{-1}\bar{\Omega}_0\bar{\Sigma}_0^{-1}\big) \quad as \quad n \to \infty,$$

*where* $\bar{\Sigma}_0 = \Sigma_1 + \Sigma_2$,

$$\Sigma_1 = g(0)E\Big(\frac{w_t}{h_t}\frac{\partial \varepsilon_t(\boldsymbol{\gamma}_0)}{\partial \boldsymbol{\theta}}\frac{\partial \varepsilon_t(\boldsymbol{\gamma}_0)}{\partial \boldsymbol{\theta}^T}\Big), \quad \Sigma_2 = \frac{1}{8}E\Big(\frac{w_t}{h_t^2}\frac{\partial h_t(\boldsymbol{\theta}_0)}{\partial \boldsymbol{\theta}}\frac{\partial h_t(\boldsymbol{\theta}_0)}{\partial \boldsymbol{\theta}^T}\Big),$$

$$\bar{\Omega}_0 = E\Big(\frac{w_t^2}{h_t}\frac{\partial \varepsilon_t(\boldsymbol{\gamma}_0)}{\partial \boldsymbol{\theta}}\frac{\partial \varepsilon_t(\boldsymbol{\gamma}_0)}{\partial \boldsymbol{\theta}^T}\Big) + \frac{E(\eta_t^2) - 1}{4}E\Big(\frac{w_t^2}{h_t^2}\frac{\partial h_t(\boldsymbol{\theta}_0)}{\partial \boldsymbol{\theta}}\frac{\partial h_t(\boldsymbol{\theta}_0)}{\partial \boldsymbol{\theta}^T}\Big).$$

*Proof.* See Theorem 2.3 of Zhu and Ling [107]. □

## 3.10 UNIT ROOT TESTS WITH INFINITE VARIANCE ERRORS

Consider the simple AR(1) model:

$$Y_t = \phi Y_{t-1} + \varepsilon_t, \tag{3.45}$$

where $Y_0$ is a constant and $\{\varepsilon_t\}$ is a sequence of identically distributed random variables with zero mean. Testing for a unit root (i.e., $H_0 : \phi = 1$) has been a long history in the literature of econometrics. A simple test is based on the asymptotic limit of the least squares estimator $\hat{\phi} = \frac{\sum_{t=1}^n Y_t Y_{t-1}}{\sum_{t=1}^n Y_{t-1}^2}$. We refer to Phillips and Perron [88] for the case when $E(\varepsilon_t^2) < \infty$ and $\{\varepsilon_t\}$ is a stationary sequence satisfying some mixing conditions. When $E(\varepsilon_t^2) = \infty$, the following theorems show that the asymptotic distribution of $n(\hat{\phi} - 1)$ under $H_0 : \phi = 1$ in (3.45) is different from the case of finite variance errors.

**Theorem 3.20.** *Assume model (3.45) satisfies that*

- $\{\varepsilon_t\}$ *is a sequence of independent and identically distributed random variables with infinite variance;*

- $P(|\varepsilon_t| > x)$ *satisfies (2.1) with* $\alpha \in (0,2)$ *and* $\lim_{x\to\infty} P(\varepsilon_t > x)/P(|\varepsilon_t| > x) = p \in [0,1]$;
- $E(\varepsilon_t) = 0$ *when* $\alpha \in (1,2)$ *and* $\varepsilon_t$ *has a symmetric distribution function around zero when* $\alpha = 1$.

*Then under* $H_0 : \phi = 1$

$$n(\hat{\phi} - 1) \xrightarrow{d} \frac{1}{2}\{U^2(1) - V(1)\}/\int_0^1 U^2(t)\,dt \quad \text{as} \quad n \to \infty,$$

*where* $(U(t), V(t))$ *is a Lévy process with the Lévy measure*

$$\tilde{\nu}(A) = \nu\{x \in \mathbb{R} : (x, x^2) \in A\}, \qquad \nu(x,\infty) = px^{-\alpha}$$
$$\text{and} \quad \nu(-\infty, -x] = (1-p)x^{-\alpha}$$

*for* $x \geq 0$.

*Proof.* See Theorem 2 of Chan and Tran [14]. □

To approximate the above asymptotic distribution, Jach and Kokoszka [61] proposed the following subsampling method.

Compute the residuals

$$\hat{\varepsilon}_t = Y_t - \hat{\phi} Y_{t-1} \quad \text{for} \quad t = 2, 3, \cdots, n$$

and the centered residuals

$$\hat{\varepsilon}_t^c = \hat{\varepsilon}_t^c - \frac{1}{n-1}\sum_{k=2}^{n} \hat{\varepsilon}_k^c \quad \text{for} \quad t = 2, \cdots, n.$$

From the centered residuals, construct $n - b$ processes of length $b$. That is, for $k = 2, \cdots, n-k+1$, define the $k$-th process as

$$y_0(k) = 0, \quad y_1(k) = \hat{\varepsilon}_k^c, \quad y_2(k) = \hat{\varepsilon}_k^c + \hat{\varepsilon}_{k+1}^c, \quad \cdots, \quad y_b(k) = \hat{\epsilon}_k^c + \cdots + \hat{\varepsilon}_{k+b-1}^c.$$

Let $\hat{\phi}_{b,k}^c$ be the least squares estimator computed from $y_1(k), \cdots, y_b(k)$. Then the following theorem can be employed to find a critical value for testing $H_0 : \phi = 1$ based on the least squares estimator $\hat{\phi}$.

**Theorem 3.21.** *Under conditions of Theorem 3.20 with* $\alpha \in (1,2)$, $H_0 : \phi = 1$, $b \to \infty$ *and* $b/n \to 0$ *as* $n \to \infty$, *we have*

$$\lim_{n\to\infty} \frac{1}{n-b}\sum_{k=2}^{n-b+1} I\big(b(\hat{\phi}_{b,k}^c - 1) \leq x\big) = P\Big(\frac{U^2(1) - V(1)}{2\int_0^1 U^2(t)\,dt} \leq x\Big) \quad \text{for all} \quad x \in \mathbb{R}.$$

*Proof.* See Theorem 3.1 of Jach and Kokoszka [61]. □

Alternatively one could employ a subsample bootstrap method to conduct such a unit root test. More specifically, draw a bootstrap sample $\varepsilon_1^*, \cdots, \varepsilon_m^*$ from $\hat{\varepsilon}_1^c, \cdots, \hat{\varepsilon}_n^c$ with replacement, and then refit the model by $Y_0^* = 0$ and $Y_t^* = \hat{\phi} Y_{t-1}^* + \varepsilon_t^*$ for $t = 1, \cdots, m$. Based on the bootstrap sample, we compute the bootstrap estimator $\hat{\phi}^* = \frac{\sum_{t=2}^m Y_{t-1}^* Y_t^*}{\sum_{t=2}^m (Y_{t-1}^*)^2}$. Then the following theorem ensures that a critical value can be obtained from repeating this procedure.

**Theorem 3.22.** *Under conditions of Theorem 3.20 with $\alpha \in (1,2)$, $H_0 : \phi = 1$, $m \to \infty$ and $m/n \to 0$ as $n \to \infty$, we have as $n \to \infty$*

$$P\big(m(\hat{\phi}^* - \hat{\phi}) \le x | \varepsilon_1, \cdots, \varepsilon_n\big) \xrightarrow{p} P\Big(\frac{U^2(1) - V(1)}{2\int_0^1 U^2(t)\,dt} \le x\Big) \text{ for all } x \in \mathbb{R}.$$

*Proof.* See Theorem 3.1 of Horváth and Kokoszka [57]. □

The next two theorems allow $\{\varepsilon_t\}$ in (3.45) to be a dependent sequence.

**Theorem 3.23.** *Assume $\varepsilon_t = \sum_{j=0}^\infty d_j e_{t-j}$ for $d_0 = 1$, $\sum_{j=0}^\infty d_j \neq 0$ and $\{e_t\}$ being a sequence of i.i.d. random variables with a symmetric distribution function lying in the normal domain of attraction of a stable law with index $\alpha \in (0,2)$. Further assume $\sum_{j=0}^\infty j|d_j|^\delta < \infty$ for some $\delta \in \big(0, \min(\alpha, 1)\big)$. Then under $H_0 : \phi = 1$*

$$n\{\hat{\phi} - 1\} \xrightarrow{d} \frac{\int_0^1 U_\alpha(t-)\,dU_\alpha(t) + \frac{1}{2}\Big(1 - \frac{\sum_{j=0}^\infty d_j^2}{(\sum_{j=0}^\infty d_j)^2}\Big)\int_0^1 \big(dU_\alpha(t)\big)^2}{\int_0^1 U_\alpha(t-)\,dU_\alpha(t)}$$

*as $n \to \infty$, where $U_\alpha(t)$ is a standard stable process with index $\alpha$ and unit scale coefficient and $U_\alpha(t-)$ denotes the left-hand limit of $U_\alpha(t)$.*

*Proof.* See Theorem 2.1 of Phillips [87]. □

**Theorem 3.24.** *Suppose model (3.45) satisfies that:*

- *$\varepsilon_t = h_t \eta_t$, $h_t^2 = \omega + \alpha h_{t-1}^2 + \beta \varepsilon_{t-1}^2$ for some $\omega > 0, \alpha \ge 0, \beta \ge 0$, where $\{\eta_t\}$ is a sequence of independent and identically distributed random variables with a symmetric distribution function;*
- *$E\{\log(\alpha + \beta \eta_t^2)\} < 0$;*
- *there exists $\kappa > 0$ such that $E\{(\alpha + \beta\eta_t^2)^\kappa\} \ge 1$ and $E\{(\alpha + \beta\eta_t^2)^\kappa \log^+(\alpha + \beta\eta_t^2)\} < \infty$;*
- *the density function of $\eta_t$ is positive in the neighborhood of zero.*

*Then there exists a unique* $\gamma \in (0, \kappa]$ *such that* $E\{(\alpha + \beta\eta_1^2)^\gamma\} = 1$ *and under* $H_0 : \phi = 1$

*i)*

$$n\{\hat{\phi} - 1\} \xrightarrow{d} \frac{\int_0^1 S_{2\gamma}(t)\, dS_{2\gamma}(t)}{\int_0^1 S_{2\gamma}^2(t)\, dt} \quad \text{for} \quad \gamma \in (0, 1) \quad \text{as} \quad n \to \infty,$$

*where* $S_{2\gamma}(t)$ *is a stable process with index* $2\gamma$*;*

*ii)*

$$n\{\hat{\phi} - 1\} \xrightarrow{d} \frac{\int_0^1 W(t)\, dW(t)}{\int_0^1 W^2(t)\, dt} \quad \text{for} \quad \gamma \geq 1 \quad \text{as} \quad n \to \infty,$$

*where* $\{W(t), 0 \leq t \leq 1\}$ *is a standard Brownian motion.*

*Proof.* See Theorems 2.1 and 2.2 in Chan and Zhang [15]. □

# CHAPTER 4

# Multivariate Regular Variation

## 4.1 MULTIVARIATE REGULAR VARIATION

As we have seen that a univariate regular variation in (2.1) provides a natural way to extrapolate the sample range of univariate data into a far tail region, one may wonder whether there is a similar way to extrapolate the sample range of multivariate data into a far tail region, which should be useful in predicting the extreme co-movement of financial markets. This chapter discusses such an extension, which is called Multivariate Regular Variation (MRV).

The distribution function of a random vector $\boldsymbol{X} = (X_1, \cdots, X_d)^T$ taking values in $\mathbb{R}^d$ is said to be a **multivariate regular variation** on the cone $\mathfrak{C} \in \mathbb{R}^d$ with a limiting measure $\mu \not\equiv 0$ if there exists a non-decreasing function $b(t) > 0$ with $\lim_{t\to\infty} b(t) = \infty$ such that

$$tP(\frac{\boldsymbol{X}}{b(t)} \in \cdot) \overset{v}{\to} \mu \quad \text{in} \quad M_+(\mathfrak{C}) \quad \text{as} \quad t \to \infty \tag{4.1}$$

in the sense of vague convergence of measures. Then condition (4.1) implies that for some non-negative constant $\alpha$

$$\mu(cB) = c^{-\alpha}\mu(B) \tag{4.2}$$

holds for all $c > 0$ and any relatively compact set $B \subset \mathfrak{C}$. That is, $\mu$ is homogeneous with order $\alpha$. Here $\alpha$ is called the tail index of the distribution function of $\boldsymbol{X}$.

From the homogeneity property (4.2), polar coordinate transformation can provide a convenient way to handle multivariate regular variation, which leads to the following equivalent definition of multivariate regular variation.

A d-dimensional random vector $\boldsymbol{X}$ is said to be regularly varying with index $\alpha \geq 0$ if there exists a sequence of constants $\{a_n\}$ and a random vector $\boldsymbol{\Theta}$ with values in $\mathbb{S}^{d-1}$, where $\mathbb{S}^{d-1}$ denotes the unit sphere in $\mathbb{R}^d$ with respect to the norm $||\cdot||$, such that for all $t > 0$,

$$nP(||\boldsymbol{X}|| > ta_n, \boldsymbol{X}/||\boldsymbol{X}|| \in \cdot) \overset{v}{\to} t^{-\alpha}P(\boldsymbol{\Theta} \in \cdot) \tag{4.3}$$

Inference for Heavy-Tailed Data.
DOI: http://dx.doi.org/10.1016/B978-0-12-804676-0.00004-3

as $n \to \infty$, where the symbol $\stackrel{v}{\to}$ stands for vague convergence on $\mathbb{S}^{d-1}$. In this case, the distribution function of $\boldsymbol{X}$ is called a multivariate regular variation, and $\alpha$ and $\boldsymbol{\Theta}$ are called the tail index and spectral measure of $\boldsymbol{X}$, respectively. In general, the norm is chosen as either the Euclidean $L_q$-norm ($|\cdot|_q$) or the max-norm ($|\cdot|_\infty$).

An equivalent expression of (4.3) is

$$\frac{P(||\boldsymbol{X}|| > tx, \boldsymbol{X}/||\boldsymbol{X}|| \in \cdot)}{P(||\boldsymbol{X}|| > t)} \stackrel{v}{\to} x^{-\alpha} P(\boldsymbol{\Theta} \in \cdot) \quad \text{as} \quad t \to \infty. \tag{4.4}$$

It follows from Hult and Lindskog [58] that $\boldsymbol{X} + \boldsymbol{c}$ for any constant $\boldsymbol{c}$ is still regularly varying with the same tail index and spectral measure as $\boldsymbol{X}$ when $\boldsymbol{X}$ is regularly varying.

Like a multivariate normal distribution, one may wonder whether a regularly varying random vector $\boldsymbol{X}$ is equivalent to that any nontrivial linear combination (i.e., $\boldsymbol{x}^T\boldsymbol{X}$ for any constant $\boldsymbol{x} \not\equiv \boldsymbol{0}$) is regularly varying, i.e.,

$$\lim_{u\to\infty} \frac{P(\boldsymbol{x}^T\boldsymbol{X} > u)}{u^{-\alpha}L(u)} = w(\boldsymbol{x}) \quad \text{for all} \quad \boldsymbol{x}, \tag{4.5}$$

where $L(u)$ is slowly varying and $w(\boldsymbol{x}) \not\equiv 0$.

The following theorem summarizes some important properties of a multivariate regular variation and answers the above question.

**Theorem 4.1.** *Let $\boldsymbol{X}$ be a random vector in $\mathbb{R}^d$.*

- *If (4.4) holds with $\alpha > 0$, then (4.5) holds with the same $\alpha$.*
- *If (4.5) holds with a positive and noninteger $\alpha$, then (4.4) holds with the same $\alpha$.*
- *If $\boldsymbol{X} \geq \boldsymbol{0}$ and (4.5) holds with an odd integer $\alpha$, then (4.4) holds with the same $\alpha$.*
- *(Multivariate version of Breiman's lemma) If the d-dimensional random vector $\boldsymbol{X}$ is regularly varying with index $\alpha \geq 0$, the d-dimensional random vector $\boldsymbol{Y}$ is independent of $\boldsymbol{X}$ and*

$$E\left\{|\boldsymbol{Y}|_2^{\alpha+\epsilon}\right\} < \infty \quad \textit{for some} \quad \epsilon > 0,$$

*then the d-dimensional random vector $\boldsymbol{XY}$ is regularly varying with the same index $\alpha$.*

*Proof.* See Basrak et al. [4]. □

**Remark 4.1.** Since (4.4) implies (4.5), one could simply apply the Hill estimator in (2.31) to estimate the tail index $\alpha$ based on some transformed data $\{\boldsymbol{\lambda}^T \boldsymbol{X}_i\}$ with $||\boldsymbol{\lambda}|| \neq 0$. See Demattes and Clémencon [29] and Kim and Lee [64].

## 4.2 HIDDEN MULTIVARIATE REGULAR VARIATION

Consider a multivariate regular variation with limit measure $\mu$ under the polar coordinate transformation. Assume that there exists a subset $B$ with $\mu(B) = 0$. Then $D := \{\mathbf{z} \in \mathfrak{C} : \frac{\mathbf{z}}{||\mathbf{z}||} \in B\}$ is a cone with zero $\mu$ measure. In this case, the limiting measure in (4.1) is degenerate in $M_+(D)$. A refined dependence structure on a subcone of $\mathfrak{C}$ can be established if we can find some new scaling function such that a new limit on the right-hand side of (4.1) is nonzero.

Let $\mathfrak{C}_0$ be a subcone in $\mathfrak{C}$. The distribution of a random vector $\mathbf{Z}$ possesses a **hidden multivariate regular variation** if, in addition to (4.1), there exists an increasing and continuous function $b^*(t) \to \infty$ such that $b(t)/b^*(t) \to \infty$ as $t \to \infty$ and there exists a Radon measure $\mu^* \not\equiv 0$ on $\mathfrak{C}_0$ such that

$$tP(\frac{\mathbf{Z}}{b^*(t)} \in \cdot) \xrightarrow{v} \mu^* \quad \text{in} \quad M_+(\mathfrak{C}_0). \tag{4.6}$$

It follows that $b^* \in RV^{\infty}_{1/\alpha^*}$ for some $\alpha^* \geq \alpha$ and

$$\mu^*(cB) = c^{-\alpha^*} \mu^*(B)$$

for some $c > 0$ and all relatively compact sets $B \subset \mathfrak{C}_0$.

## 4.3 TAIL DEPENDENCE AND EXTREME RISKS UNDER MULTIVARIATE REGULAR VARIATION

Let $\boldsymbol{Z} = (Z_1, \cdots, Z_d)^T$ be a nonnegative random vector with distribution function $F$ and continuous marginals $F_j$ for $j = 1, \cdots, d$. To model dependence efficiently and flexibly, the so-called copula and survival copula of $\boldsymbol{Z}$ become popular in insurance and finance as recommended in Basel III for banking industry and Solvency II for insurance business, which are respectively defined as

$$C(u_1, \cdots, u_d) = P(F_1(Z_1) \leq u_1, \cdots, F_d(Z_d) \leq u_d) = F(F_1^-(u_1), \cdots, F_d^-(u_d))$$

and

$$\bar{C}(u_1, \cdots, u_d) = P(F_1(Z_1) > 1 - u_1, \cdots, F_d(Z_d) > 1 - u_d)$$

for $0 \le u_1, \cdots, u_d \le 1$. For modeling/predicting extreme risks, the so-called lower and upper tail dependence functions of $\boldsymbol{Z}$ are useful and are respectively defined as

$$b(\boldsymbol{w}) = \lim_{t \to 0} \frac{C(t\boldsymbol{w})}{t} \quad \text{and} \quad b^*(\boldsymbol{w}) = \lim_{t \to 0} \frac{\bar{C}(t\boldsymbol{w})}{t} \quad \text{for} \quad \boldsymbol{w} \ge \boldsymbol{0} \quad \text{and} \quad \boldsymbol{w} \ne \boldsymbol{0}.$$

By focusing on the upper tail dependence and assuming that (4.1) holds for $\boldsymbol{Z}$ and the marginal distribution functions are tail equivalent, i.e.,

$$\lim_{t \to \infty} \frac{1 - F_i(t)}{1 - F_j(t)} = 1 \quad \text{for} \quad i \ne j, \tag{4.7}$$

we study the extreme behavior of two commonly employed risk measures, Value-at-Risk and Expected Shortfall, in industry, which are respectively defined as

$$VaR_p(|\boldsymbol{Z}|_1) = \inf\{t : P(|\boldsymbol{Z}|_1 > t) \le 1 - p\}$$

and

$$ES_p(|\boldsymbol{Z}|_1) = E\big(|\boldsymbol{Z}|_1 | \boldsymbol{Z} > VaR_p(|\boldsymbol{Z}|_1)\big).$$

Note that $\boldsymbol{Z} \ge \boldsymbol{0}$ implies that $|\boldsymbol{Z}|_1 = \sum_{i=1}^d |Z_i| = \sum_{i=1}^d Z_i$.

**Theorem 4.2.** *Suppose (4.1), (4.7) hold and $1 - F_1$ is a regular variation with index $-\beta$ at infinity.*

*i) The upper tail dependence function $b^*(\cdot)$ exists and*

$$b^*(\boldsymbol{w}) = \frac{\mu([\boldsymbol{w}^{-1/\beta}, \infty])}{\mu([1, \infty] \times \bar{\mathbb{R}}_+^{d-1})}, \quad \textit{where} \quad \bar{\mathbb{R}}_+ = [0, \infty].$$

*ii) If $F$ is absolutely continuous and the partial derivative $\partial^d b^*(\boldsymbol{v})/\partial v_1 \cdots \partial v_d$ exists everywhere, then*

$$\begin{aligned} q(\beta, b^*) &:= \lim_{t \to \infty} \frac{P(|\boldsymbol{Z}|_1 > t)}{1 - F_1(t)} = (-1)^d \int_W \frac{\partial^d b^*(\boldsymbol{w}^{-\beta})}{\partial w_1 \cdots \partial w_d} \, d\boldsymbol{w} \\ &= \int_{W^{-1/\beta}} \frac{\partial^d b^*(\boldsymbol{v})}{\partial v_1 \cdots \partial v_d} \, d\boldsymbol{v}, \end{aligned}$$

*where* $W = \{\boldsymbol{w} = (w_1, \cdots, w_d)^T : |\boldsymbol{w}|_1 > 1\}$ *and* $W^{-1/\beta} = \{\boldsymbol{v} = (v_1, \cdots, v_d)^T : |\boldsymbol{v}^{-1/\beta}| > 1\}$.

*iii) If the tail dependence function* $b^*(\mathbf{1}) > 0$, *then*

$$\lim_{p \to 1} \frac{VaR_p(|\boldsymbol{Z}|_1)}{VaR_p(Z_1)} = \left\{q(\beta, b^*)\right\}^{1/\beta}.$$

*iv)*

$$\lim_{p \to 1} \frac{ES_p(|\boldsymbol{Z}|_1)}{VaR_p(|\boldsymbol{Z}|_1)} = \sum_{j=1}^{d} u_j(B; \mu),$$

*where* $B = \{\boldsymbol{x} = (x_1, \cdots, x_d)^T : |\boldsymbol{x}|_1 > 1\}$ *and*

$$u_j(B; \mu) = \int_0^\infty \frac{\mu(A_j(w) \cap B)}{\mu(B)} dw$$

*with* $A_j(w) = \{\boldsymbol{x} = (x_1, \cdots, x_d)^T : x_j > w\}$ *for* $j = 1, \cdots, d$.

*Proof.* See Joe and Li [63] and Weng and Zhang [104]. □

## 4.4 LOSS GIVEN DEFAULT UNDER MULTIVARIATE REGULAR VARIATION

Consider a portfolio with $n$ obligors and net losses $X_i$ for each obligor. The $i$-th obligor defaults once the loss $X_i$ is larger than some deterministic threshold $a_i > 0$. For each $i = 1, \cdots, n$, we have a loss settlement function $G_i$ satisfying that

$$G_i \text{ is nondecreasing with } G_i(s) = 0 \text{ for } s \leq 0 \text{ and } G_i(\infty) = 1. \tag{4.8}$$

Then the loss given default is defined as

$$L = \sum_{i=1}^{n} e_i G_i(\frac{X_i}{a_i} - 1), \tag{4.9}$$

where $e_1, \cdots, e_n$ are $n$ positive constants denoting scaled exposures such that $\sum_{i=1}^{n} e_i = 1$. To investigate the asymptotic behavior of $L$ near the upper endpoint, Tang and Yuan [101] proposed to study the extreme properties of $K = (1 - L)^{-1}$ under the following conditions.

**A1)** The random vector $\boldsymbol{X} = (X_1, \cdots, X_n)^T$ satisfies (4.1), (4.2) and $c_i = \mu((\mathbf{1}_i, \infty]) > 0$ for $i = 1, \cdots, n$, where $\mathbf{1}_i$ denotes the $n$-dimensional vector with the $i$-th element being one and the rest being zero.

**A2)** There is a distribution function $G$ concentrated on $[0, \infty)$ with $\bar{G} \in RV_{-\beta}^{\infty}$ for some $\beta \in (0, \infty)$ such that

$$\lim_{x\to\infty} \frac{\bar{G}_i(x)}{\bar{G}(x)} = d_i > 0 \quad \text{for} \quad i = 1, \cdots, n.$$

**Theorem 4.3.** *Assume conditions A1), A2), (4.8) and $\sum_{i=1}^n e_i = 1$ hold. Put $H(x) = 1/b^-(x)$ with $b(x)$ given in (4.1).*

*i) We have*

$$\lim_{x\to\infty} \frac{P(K > x)}{H(G^-(1 - 1/x))} = \tilde{\mu}(A),$$

*where $A = \{\boldsymbol{y} \in [0, \infty]^n : \sum_{i=1}^n e_i/y_i < 1\}$ and $\tilde{\mu}$ is a Radon measure on $[0, \infty]^n \backslash \{\mathbf{0}\}$ defined by*

$$\tilde{\mu}([\mathbf{0}, \boldsymbol{t}]^c) = \mu([\mathbf{0}, (a_1(d_1 t_1)^{1/\beta}, \cdots, a_n(d_n t_n)^{1/\beta})^T]^c)$$

*for $\boldsymbol{t} = (t_1, \cdots, t_n)^T > \mathbf{0}$, where $D^c$ denotes the complementary set of $D$.*

*ii) If further $\alpha/\beta > 1$ and $\mu((\mathbf{1}, \boldsymbol{\infty}]) > 0$, then*

$$\lim_{p\to 1} \frac{VaR_p(K)}{1/\bar{G}((1 - H)^-(p))} = \{\tilde{\mu}(A)\}^{\beta/\alpha}$$

$$\text{and} \quad \lim_{p\to 1} \frac{ES_p(K)}{1/\bar{G}((1 - H)^-(p))} = \frac{\alpha\,\{\tilde{\mu}(A)\}^{\beta/\alpha}}{\alpha - \beta}.$$

*Proof.* See Tang and Yuan [101]. □

When the loss in (4.9) is modeled by a latent vector, Shi et al. [97] defined a new loss given default as

$$L^*(\boldsymbol{p}) = \sum_{i=1}^n e_i G_i\left(\frac{X_i}{F_i^-(1 - p_i)} - 1\right) \quad \text{for} \quad \boldsymbol{p} = (p_1, \cdots, p_n)^T > \mathbf{0}. \tag{4.10}$$

**Theorem 4.4.** *Assume A1), (4.8), $\sum_{i=1}^n e_i = 1$ and $\lim_{p\to 0} p_i/p = b_i > 0$ for $i = 1, \cdots, n$. Then*

$$\lim_{p\to 0} \frac{P(L^*(\boldsymbol{p}) > l)}{p} = \mu(A_l),$$

*where*

$$A_l = \left\{\boldsymbol{x} \in [0, \infty]^n : \sum_{i=1}^n e_i G_i((c_i/b_i)^{-1/\alpha} x_i - 1) > l\right\}.$$

*Proof.* See Shi et al. [97]. □

## 4.5 ESTIMATING AN EXTREME SET UNDER MULTIVARIATE REGULAR VARIATION

Let $\{\mathbf{X}_i = (X_{i,1}, \cdots, X_{i,d})^T\}_{i=1}^n$ be independent and identically distributed random vectors with common distribution function $F$ and density function $f$. Consider an extreme region

$$Q_n = \left\{ \boldsymbol{z} \in \mathbb{R}^d : f(\boldsymbol{z}) \leq \beta_n \right\},$$

where $\beta_n$ is such that $P(Q_n) = p_n > 0$ and $p_n \to 0$ as $n \to \infty$. When $p_n$ is too small, nonparametric estimation of the set $Q_n$ fails, and some assumptions on the tail are needed. Cai et al. [12] proposed the following method to estimate the extreme set $Q_n$ by assuming that the density is a multivariate regular variation.

**B1)** Suppose there exist a positive function $q$ and a positive function $V$ regularly varying at infinity with negative index $-\alpha$ such that

$$\lim_{t\to\infty} \frac{f(t\boldsymbol{z})}{t^{-d}V(t)} = q(\boldsymbol{z}) \quad \text{for all} \quad \boldsymbol{z} \neq 0$$

and

$$\lim_{t\to\infty} \sup_{|\boldsymbol{z}|_2=1} \left|\frac{f(t\boldsymbol{z})}{t^{-d}V(t)} - q(\boldsymbol{z})\right| = 0.$$

Without loss of generality, we can choose

$$V(t) = P(|\mathbf{X}_1|_2 > t),$$

and then (4.1) holds with $\mu(B) = \int_B q(\boldsymbol{z})\, d\boldsymbol{z}$.

The key idea of estimating $P(Q_n)$ is to connect $Q_n$ to the fixed set $S = \{\boldsymbol{z} : q(\boldsymbol{z}) \leq 1\}$ by using condition B1). Put $H(s) = 1 - V(s)$ and $U(t) = H^{-}(1 - 1/t)$. First we approximate the set $Q_n$ by

$$\bar{Q}_n = \left\{ \boldsymbol{z} : f(\boldsymbol{z}) \leq \left(\frac{np}{k\mu(S)}\right)^{1+d/\alpha} \frac{k/n}{(U(n/k))^d} \right\},$$

where

**B2)** $k = k_n$ is a sequence of positive integers such that $k < n$, $k \to \infty$ and $k/n \to 0$ as $n \to \infty$.

Next, $\bar{Q}_n$ is approximated by another set

$$\tilde{Q}_n = U\left(\frac{n}{k}\right)\left(\frac{k\mu(S)}{np}\right)^{1/\alpha} S,$$

which is independent of $f$, but depends on the limiting function $q$ since $S = \{z : q(z) \leq 1\}$ is determined by $q$. Put

$$B_{r,A} = \left\{z \in \mathbb{R}^d : |z|_2 \geq r, z/|z|_2 \in A\right\} \quad \text{and} \quad \Psi(A) = \mu(B_{1,A})$$

for any Borel set $A$ on $\Theta = \{z \in \mathbb{R}^d : |z|_2 = 1\}$. The existence of a density $q$ of $\mu$ implies the existence of a density $\psi$ of $\Psi$ such that

$$\Psi(A) = \int_A \psi(w)\, d\lambda(w),$$

where $\lambda$ is the Hausdorff measure on $\Theta$ and

$$q(rw) = \alpha r^{-\alpha-d}\psi(w).$$

Next we write $S$ and $\mu(S)$ in terms of the spectral density as

$$S = \left\{z = rw : r \geq (\alpha\psi(w))^{1/(\alpha+d)}, w \in \Theta\right\}$$

and

$$\mu(S) = \alpha^{-\alpha/(\alpha+d)} \int_\Theta (\psi(w))^{d/(\alpha+d)}\, d\lambda(w).$$

Finally, to estimate $\tilde{Q}_n$, we need estimators for $U(n/k), \alpha, S, \mu(S)$.

Put $R_i = |\mathbf{X}_i|_2$, $\boldsymbol{W}_i = \mathbf{X}_i/R_i$ and let $R_{n,1} \leq \cdots \leq R_{n,n}$ denote the order statistics of $R_1, \cdots, R_n$. Let $K : [0,1] \to [0,1]$ be a continuous and nonincreasing function with $K(0) = 1$ and $K(1) = 0$, and let $\hat{\alpha}$ denote a tail index estimator of $\alpha$, say the Hill estimator in (2.31). Then for $h \in (0,1)$, define

$$\hat{U}(\frac{n}{k}) = R_{n,n-k}, \; \hat{\Psi}_n(w) = \frac{c(h,K)}{k}\sum_{i=1}^n K(\frac{1 - w^T \boldsymbol{W}_i}{h}) I(R_i > R_{n,n-k}),$$

$$\frac{1}{c(h,K)} = \int_{C_w(h)} K(\frac{1 - v^T w}{h})\, d\lambda(v), \; C_w(h) = \left\{v \in \Theta : w^T v \geq 1 - h\right\},$$

$$\hat{S} = \left\{z = rw : r \geq (\hat{\alpha}\hat{W}_n(w))^{\frac{1}{\hat{\alpha}+d}}, w \in \Theta\right\}, \; \widehat{\mu(S)} = \hat{\alpha}^{-\frac{\hat{\alpha}}{\hat{\alpha}+d}} \int_\Theta (\hat{\Psi}_n(w))^{\frac{1}{\hat{\alpha}+d}}\, d\lambda(w),$$

and finally the proposed estimator of $Q_n$ is

$$\hat{Q}_n = \hat{U}(\frac{n}{k})(\frac{k\widehat{\mu(S)}}{np})^{1/\hat{\alpha}}\hat{S}.$$

To derive the consistency of the above proposed estimator $\hat{Q}_n$, Cai et al. [12] further assumed the following conditions.

**B3)** $\lim_{t\to\infty} \frac{U(t)}{t^{1/\alpha}} = c \in (0, \infty)$.

**B4)** $\sqrt{k}(\hat{\alpha} - \alpha) = O_p(1)$ as $n \to \infty$.

**B5)** $\frac{\log(np_n)}{\sqrt{k}} \to 0$, $h = h(n) \to 0$, $\frac{k}{c(h,K)\log k} \to \infty$, $p_n \to 0$ as $n \to \infty$.

**Theorem 4.5.** *Under conditions B1)–B5),*

$$\frac{P(\hat{Q}_n \Delta Q_n)}{P(Q_n)} \xrightarrow{p} 0 \quad \textit{and} \quad \frac{P(\hat{Q}_n)}{P(Q_n)} \xrightarrow{p} 1 \quad \textit{as} \quad n \to \infty,$$

*where $\Delta$ means the symmetric difference of two sets.*

*Proof.* See Cai et al. [12]. □

## 4.6 EXTREME GEOMETRIC QUANTILES UNDER MULTIVARIATE REGULAR VARIATION

Let $\boldsymbol{Z} = (Z_1, \cdots, Z_d)^T$ with $d \geq 2$ be a random vector. For a given vector $\boldsymbol{v}$ in the unit open ball $\mathbb{B}^d$ of $\mathbb{R}^d$, a geometric quantile related to $\boldsymbol{v}$ is defined as

$$\arg\min_{\boldsymbol{q}\in\mathbb{R}^d} \left\{ E(|\boldsymbol{X} - \boldsymbol{q}|_2 - |\boldsymbol{X}|_2) - \boldsymbol{v}^T\boldsymbol{q} \right\}.$$

When $\boldsymbol{X}$ has a density function $f$, the above minimization can be written as

$$\arg\min_{\boldsymbol{q}\in\mathbb{R}^d} \left\{ \int_{\mathbb{R}^d} (|\boldsymbol{x} - \boldsymbol{q}|_2 - |\boldsymbol{x}|_2) f(\boldsymbol{x})\, d\boldsymbol{x} - \boldsymbol{v}^T\boldsymbol{q} \right\}.$$

Let $\mathbb{S}^{d-1}$ denote the unit hypersphere of $\mathbb{R}^d$. For any $\boldsymbol{u} \in \mathbb{S}^{d-1}$, let $\Pi_{\boldsymbol{u}} : \boldsymbol{y} \to \boldsymbol{y} - \boldsymbol{y}^T\boldsymbol{u}\boldsymbol{u}$ denote the orthogonal projection onto the hyperplane of $\mathbb{R}^d$. For any positive definite $d \times d$ symmetric matrix $\Sigma$, we define the ellipsoid

$$E_{\Sigma}^{d-1} = \{\boldsymbol{x} \in \mathbb{R}^d : \boldsymbol{x}^T \Sigma^{-1} \boldsymbol{x} = 1\}$$

and its related surface measure $\mu_\Sigma$ given by

$$\mu_\Sigma(C) = (\det \Sigma)^{1/2} \sigma(\Sigma^{-1/2} C)$$

for every Borel measurable subset $C$ of $E_{\Sigma}^{d-1}$, where $\sigma$ is the standard surface measure on $\mathbb{S}^{d-1}$. Here $\det \Sigma$ denotes the determinant of $\Sigma$.

Girard and Stupfler [41] derived the following asymptotic results for $\boldsymbol{q}(\lambda\boldsymbol{u})$ as $\lambda \to 1$ by assuming that

**($\mathbf{M}_\alpha$)** The density function $f$ of $\boldsymbol{X}$ is a continuous function on a neighborhood of infinity, such that the function $\boldsymbol{y} \to |\boldsymbol{y}|_2^d f(\boldsymbol{y})$ is bounded in any compact neighborhood of zero and there exist a positive function Q on $\mathbb{R}^d$ and a function $V$ which is regularly varying at infinity with index $-\alpha < 0$ such that

$$\lim_{t\to\infty} \frac{f(t\boldsymbol{y})}{t^{-d}V(t)} = Q(\boldsymbol{y}) \quad \text{for} \quad \boldsymbol{y} \neq \boldsymbol{0}$$

$$\text{and} \quad \lim_{t\to\infty} \sup_{\boldsymbol{w}\in\mathbb{S}^{d-1}} \left|\frac{f(t\boldsymbol{w})}{t^{-d}V(t)} - Q(\boldsymbol{w})\right| = 0.$$

**Theorem 4.6.** *Let $\boldsymbol{u} \in \mathbb{S}^{d-1}$ and $\Sigma$ be an arbitrary positive $d \times d$ symmetric matrix.*

- *If ($M_\alpha$) holds with $\alpha \in (0,1)$, then as $\lambda \to 1$*

$$\frac{1}{V(|\boldsymbol{q}(\lambda\boldsymbol{u})|_2)}\left(\frac{\boldsymbol{q}(\lambda\boldsymbol{u})}{|\boldsymbol{q}(\lambda\boldsymbol{u})|_2} - \boldsymbol{u}\right) \to \int_{\mathbb{R}^d} \frac{\Pi_{\boldsymbol{u}}(\boldsymbol{y})}{|\boldsymbol{y}-\boldsymbol{u}|_2} Q(\boldsymbol{y})\, d\boldsymbol{y}.$$

- *If ($M_\alpha$) holds with $\alpha \in (0,2)$, then as $\lambda \to 1$*

$$\frac{1-\lambda}{V(|\boldsymbol{q}(\lambda\boldsymbol{u})|_2)} \to \int_{\mathbb{R}^d} \left(1 + \frac{(\boldsymbol{y}-\boldsymbol{u})^T\boldsymbol{u}}{|\boldsymbol{y}-\boldsymbol{u}|_2}\right) Q(\boldsymbol{y})\, d\boldsymbol{y}.$$

- *If ($M_1$) holds and $\lim_{t\to\infty}\int_1^t r^{\alpha-1}V(r)\,dr = \infty$, then as $\lambda \to 1$*

$$\frac{|\boldsymbol{q}(\lambda\boldsymbol{u})|_2}{\int_1^{|\boldsymbol{q}(\lambda\boldsymbol{u})|_2} r^{\alpha-1}V(r)\,dr}\left(\frac{\boldsymbol{q}(\lambda\boldsymbol{u})}{|\boldsymbol{q}(\lambda\boldsymbol{u})|_2} - \boldsymbol{u}\right) \to \int_{E_\Sigma^{d-1}} \Pi_{\boldsymbol{u}}(\boldsymbol{w}) Q(\boldsymbol{w}) \mu_\Sigma(d\boldsymbol{w}).$$

- *If ($M_2$) holds and $\lim_{t\to\infty}\int_1^t r^{\alpha-1}V(r)\,dr = \infty$, then as $\lambda \to 1$*

$$\frac{|\boldsymbol{q}(\lambda\boldsymbol{u})|_2^2}{\int_1^{|\boldsymbol{q}(\lambda\boldsymbol{u})|_2} r^{\alpha-1}V(r)\,dr}(1-\lambda) \to \frac{1}{2}\int_{E_\Sigma^{d-1}} (\Pi_{\boldsymbol{u}}(\boldsymbol{w}))^T \boldsymbol{w} Q(\boldsymbol{w}) \mu_\Sigma(d\boldsymbol{w}).$$

*Proof.* See Theorems 1 and 2 of Girard and Stupfler [41]. □

# CHAPTER 5

# Applications

After introducing some visualization techniques to examine the tail behavior of data, this chapter illustrates how to combine these visualization tools with the developed procedures in previous chapters to analyze some data sets in insurance and finance.

## 5.1 SOME VISUALIZATION TOOLS FOR PRELIMINARY ANALYSIS

### 5.1.1 Hill Plot

Let $X_1, \cdots, X_n$ be a random sample from distribution function $F$. We assume $P(X_i > 0) = 1$. Define the Hill estimator $\hat{\alpha}(k)$ in (2.31). Mason [72] showed that the following statements are equivalent:

**A.** (2.1) holds;
**B.** There exists a $\theta \in (0, 1)$ such that $\hat{\alpha}([n^\theta]) \to \alpha$ in probability;
**C.** There exists a $\theta \in (0, 1)$ such that $\hat{\alpha}([n^\theta]) \to \alpha$ almost surely;
**D.** $\hat{\alpha}(k) \to \alpha$ in probability whenever $k = k(n)$ satisfies condition (2.28).

Therefore, the consistency of the Hill estimator also indicates the regularity of the distribution where the sample comes from. The so-called Hill plot, plot of $(k, \hat{\alpha}(k))$, $1 \le k < n$, can be used to verify condition (2.1) graphically if the sample size is sufficiently large. If the Hill plot is relatively stable in a range of $k$-values from small to moderately large, it may serve as an evidence for having a heavy tailed distribution in a preliminary analysis.

### 5.1.2 Alternative Hill Plot

When the Hill plot is not very informative, Resnick and Stărică [94] suggested a simple alternative plot of the Hill estimator (called alt-Hill plot below). More specifically, the alt-Hill plot is constructed by plotting $\left(\theta, \hat{\alpha}_{[n^\theta]}\right)$, $\theta \in (0, 1)$. This is similar to the Hill plot when one uses a logarithmic scale for the $k$-axis. It has the effect of stretching the left half of the Hill plot and giving more display space to smaller values of $k$.

As pointed out in Drees et al. [34], the Hill plot is preferred when the underlying distribution is a Pareto distribution, but the alt-Hill plot is useful in a wide variety of circumstances such that the underlying distribution

Inference for Heavy-Tailed Data.
DOI: http://dx.doi.org/10.1016/B978-0-12-804676-0.00005-5

satisfies the second-order regular variation condition (2.4). A general recommendation is to produce both the Hill plot and the alt-Hill plot and compare them to identify a good sample fraction.

### 5.1.3 Log-Quantile Plot

Recall that a heavy-tailed distribution $F$ satisfies condition (2.2) (or equivalently (2.1)), and one always has the expression $\bar{F}^{-}(x) = x^{-1/\alpha}L(x)$, $x > 0$, where $L(x) > 0$ is a slowly varying function at zero. From Potter's bound or representation theorem of a regular variation we have $\log L(x) = o(\log(x^{-1}))$ as $x \to 0$, which implies

$$\lim_{x\to 0} \frac{\log \bar{F}^{-}(x)}{\log(x^{-1})} = \frac{1}{\alpha}.$$

Indeed by using Potter's bound we can show that $X_{n,n-i+1}/\bar{F}^{-}(i/(n+1))$ is bounded away from 0 and infinity in probability for $1 \le i \le k$, where $k = k(n)$ is a sequence of integers satisfying condition (2.28). This implies that $\log X_{n,n-i+1}/\log((n+1)/i)$ is approximately equal to $1/\alpha$ for $1 \le i \le k$. When we plot the log-quantiles $(-\log(1-i/(n+1)), \log X_{n,i})$, $1 \le i \le n$, the upper part of the plot should form a relatively straight line if $F$ is a heavy-tailed distribution. More precisely, $(-\log(1-i/(n+1)), \log X_{n,i})$, $n-k+1 \le i \le n$ should lie around a relatively straight line.

We conduct a simulation study to illustrate the above three visualization techniques. The first random sample of size 1000 is drawn from Pareto distribution with $\alpha = 1$. The second sample is generated by simply adding a constant 5 to the first one. The Hill plot, alt-Hill plot and log-quantile plot based on the first sample and the second sample are given in the left panel and right panel, respectively, of Fig. 5.1. These plots confirm the discovery in Drees et al. [34], that is, for the Pareto distribution, the Hill plot is superior to the alt-Hill plot, while for the second distribution which is heavy-tailed with a second-order term, the alt-Hill plot is more effective. There is no significant difference between the log-quantile plots from the two samples, and both illustrate the linearity on the far right end.

## 5.2 HEURISTIC APPROACH FOR TRAINING DATA

Before we analyze some real data sets, we first examine some simulated data with a heavy-tailed distribution. We choose Pareto and Burr distributions defined in Section 2.1 as examples in our study.

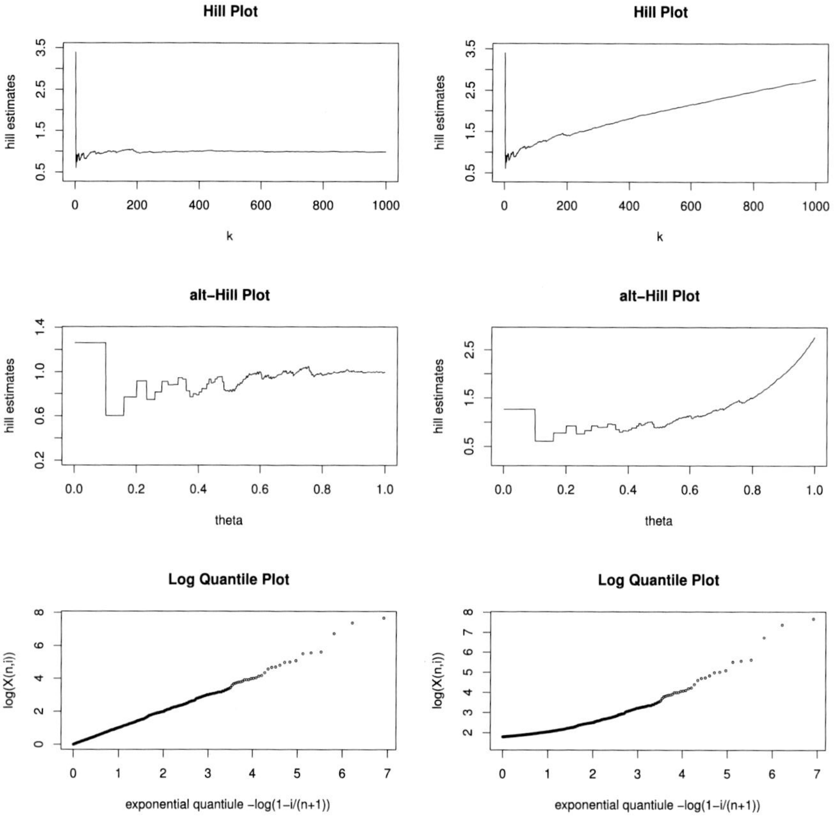


**Figure 5.1 Comparison of three visualization tools** based on a random sample of size 1000. The three plots on the left panel are generated from random variable $X$ with Pareto (1) distribution, and the three plots on the right panel are from random variable $X + 5$. The value of the tail index is $\alpha = 1$.

- A Pareto($\alpha$) distribution has a tail index $\alpha$, with mean $\frac{\alpha}{\alpha-1}$ if $\alpha > 1$.
- A Burr($a, b$) distribution has a tail index $ab$, with mean $a\beta(1 + 1/b, a - 1/b)$ if $ab > 1$, where $\beta(c, d) = \int_0^1 t^{c-1}(1-t)^{d-1}\,dt$ is the Beta function for $c, d > 0$.

A simulation study is conducted by using the statistical software **R** and random number generators "rpareto" and "rburr" from **R** package "actuar" for Pareto and Burr distributions. For all simulations we use a fixed seed 123. Below are two examples to get a random sample of 1000 observations from Pareto(2) and Burr(4, 2):

```
set.seed(123)
x=rpareto(1000, 2, 1)+1

set.seed(123)
y=rburr(1000, 4, 2)
```

We summarize the following observations from the theoretical properties of the Hill estimator and the conducted simulation study.

1. Assume that a random variable $X$ has a heavy-tailed distribution with index $\alpha$. Then under a power transformation, a scale transformation and a location shift, the new random variables are also heavy-tailed, that is, the distributions of $X^a$, $aX$ and $X+b$ are heavy-tailed with tail indices $\alpha/a$, $\alpha$, and $\alpha$, respectively for any numbers $a>0$ and $b\in\mathbb{R}$. The Hill estimator is invariant under a scale transformation, and a power transformation will result in a new Hill estimator proportional to the Hill estimator computed from the original observations. However, a small shift in location may result in quite different estimates from the Hill plot, see, e.g, the Hill plots in Fig. 5.1. Sometimes, a shift in location may improve the performance of the Hill estimator; see item 4 below for more details.
2. The ideal model for the Hill estimator is the Pareto distributions, Pareto($\alpha$), where condition (2.1) holds trivially since we have $\bar{F}(tx)/\bar{F}(t)=x^{-\alpha}$ for any $x>0$ if $t$ is sufficiently large. This is equivalent to a second order regular variation condition with $\rho=\infty$. In this case, the following central limit theorem for the Hill estimator holds:

$$\sqrt{k}\left\{\hat{\alpha}(k)-\alpha\right\}\xrightarrow{d}N(0,\alpha)$$

as long as $k=k(n)<n$ and $k\to\infty$ as $n\to\infty$.
3. If the second order regular variation condition (2.4) holds, the performance of the Hill estimator depends on the second order regular variation parameters. The Hill estimator improves when the sample size increases; see Fig. 5.2 for the Hill plots for samples taken from Burr(2, 0.75) distribution. We also note that the estimation error for the Hill estimator gets smaller when the sample size $n$ increases.
4. Extreme values from some heavy-tailed distributions may not be large at all if the sample size is not truly large, especially when the tail index is larger. For example, for a Burr(3, 1) distribution, its 99.9% and 99.99% quantiles are about 9.00 and 20.54 , respectively. In this case, a smaller shift in location can change the Hill estimator significantly. To illustrate

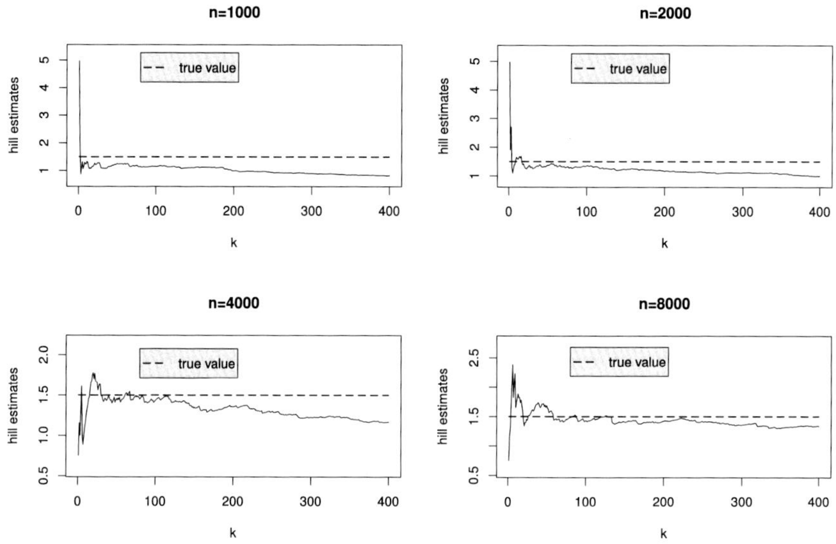


**Figure 5.2 Hill plots** for samples with different sizes from Burr(2, 0.75) distribution ($n =$ 1000, 2000, 4000, and 8000).

this, we generate a random sample of size $n = 1000$ from Burr(3, 1) distribution, and then we add 1 to this sample and get a new sample. Two Hill plots are given in Fig. 5.3. We observe from the two Hill plots that the Hill estimates based on the original observations perform very poorly in general while the Hill estimates based on the transformed data are much better. This phenomenon is not surprising as we can show that for a random variable $X$ with a Burr($a$, 1) for any $a > 0$, the second order term (i.e. $A(t)$ in (2.4)) for the distribution of $X + 1$ converges to zero at a much faster rate than that for the distribution of $X$.

**5.** Under the second order regular variation condition, if the sample fraction $k$ involved in the estimation increases, the bias will gradually dominate the estimator. Since the second order regular variation function $A(t)$ in (2.4) has a constant sign ultimately, the Hill plot will clearly show either an upward trend or a downward trend. For statistical inference such as constructing confidence intervals and testing hypotheses, it is important to decide a range for values of the sample fraction $k$ such that the effect of bias of the estimator is as small as possible. In the ideal case like the Hill plot (left panel) in Fig. 5.1 and the Hill plot (right panel) in Fig. 5.3, one can use a value of $k$ as large as $n - 1$. In general, the feasible range for $k$ is very limited as condition (2.28) should be sat-

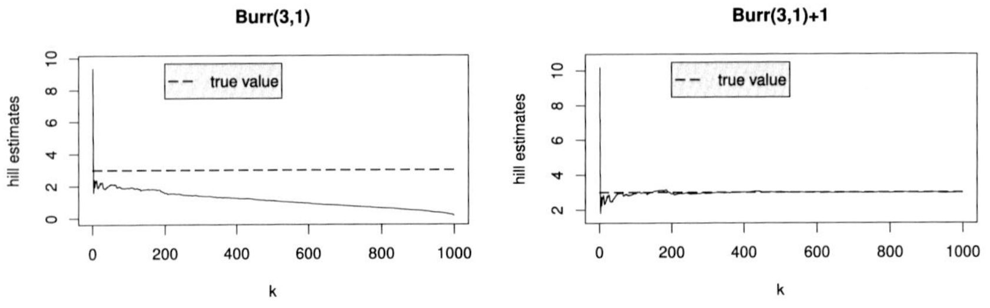


**Figure 5.3** **Hill plots** for a sample of size $n = 1000$ from Burr(3, 1) distribution and its transformation with a shift 1 in location.

isfied. We list some common features on the Hill plots for heavy-tailed data as follows.

- The Hill estimate with very small $k$ fluctuates and is not very reliable for inference. Though, an average of a few Hill estimates over some small values of $k$ can be very informative sometimes.
- There should be a range for relatively small values of $k$ where the Hill estimates are relatively stable or fluctuate slightly. As the sample fraction $k$ increases, estimation bias will become non-negligible, and those values of $k$ will form a range where the Hill estimates form a downward or an upward trend.
- We can identify a value of $k$ that can be regarded as a turning point at which the Hill estimates start to exhibit a trend. The Hill estimate using this sample fraction usually gives a satisfactory solution for estimating the tail index.

**Example 5.1.** We consider the random sample of size 1000 which has been used for the Hill plot in Fig. 5.2. From the Hill plot, we see that the Hill estimate decreases very slowly when the sample fraction $k$ increases. We shouldn't use the Hill estimates with very small values of $k$. We notice that the Hill estimates for $k = 1, 2$ are much larger than other estimates. The maximum values of the Hill estimates for $k \geq 3$ is 1.328. Recall that the tail index of Burr(2, 0.75) distribution is 1.5. There seems a quite large difference between the true value of the tail index and its estimates. In order to understand how to choose $k$, we perform the following analyses.

- We first run the six goodness-of-fit tests in Chapter 2 and get a range of values of the sample fraction $k$ for detecting a heavy tail. In Fig. 5.4, the BJI test statistics are plotted and the test is not significant at level 0.05 in the range from 2 to 185 except 132. The other five tests are

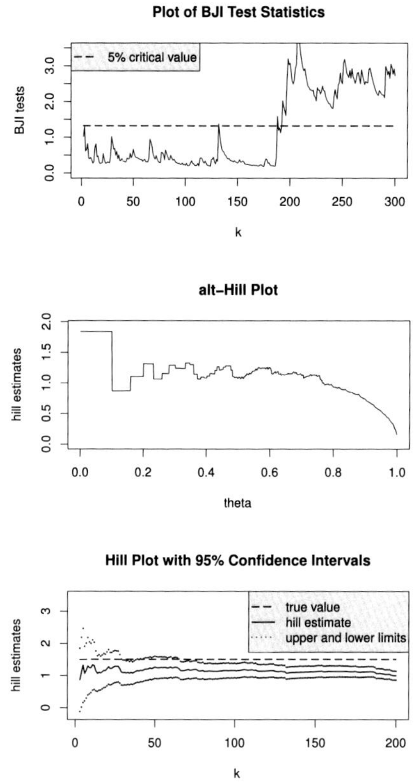


**Figure 5.4 BJI test, alt-Hill plot and Hill plot** for confidence intervals based on a sample of size $n = 1000$ from Burr(2, 0.75) distribution.

not significant in the range from 2 to 185. So we can choose a sample fraction $k$ from 2 to 131.

- From the alt-Hill plot in Fig. 5.4, we can see a clear turning point occurs near $\theta = 0.6$, which corresponds to the sample fraction $k \approx 1000^{0.6} = 63.09$. Hence we prefer to take $k = 63$.
- With the selected $k = 63$, the Hill estimate for $\alpha$ is $\hat{\alpha}(63) = 1.257$. We can assess this estimate by constructing a 95% confidence interval, which is $(0.946, 1.567)$. Hence the tail index $\alpha$ for Burr(2, 0.75), 1.5, does fall into this interval.
- Moreover, we plot 95% confidence intervals of the tail index $\alpha$ for all $k$ in the range from 3 to 200 in Fig. 5.4. The plot for the confidence limits clearly indicates that the performance of the Hill estimate is very poor when $k$ is large, even for those values of $k$ lightly larger than 63.

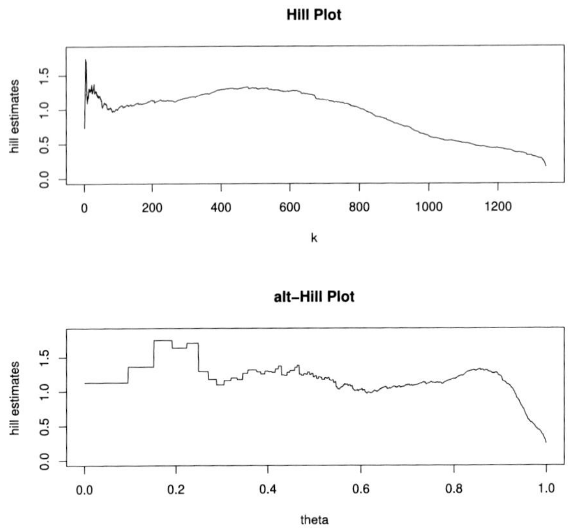


**Figure 5.5 Plots for automobile bodily injury claims.**

## 5.3 APPLICATIONS TO INDEPENDENT DATA

In practice, analysis of the tail behavior of a distribution function can be very complicated. Due to the nature of a heavy-tailed distribution, frequently a very large sample size is required for a sample to exhibit the tail property of the underlying distribution.

We study several data sets below and try to answer the following questions (if applicable):

- Is the underlying distribution heavy-tailed?
- What is the tail index?
- How to estimate the mean and/or quantiles of the underlying distribution?

### 5.3.1 Automobile Bodily Injury Claims

**Data description.** The automobile injury claims dataset is from the Insurance Research Council (IRC), a division of the American Institute for Chartered Property Casualty Underwriters and the Insurance Institute of America. The dataset, collected in 2002, includes demographic information about the claimant, attorney involvement, the economic loss (in thousands), and other variables. The dataset used here is a sample of $n = 1340$ losses

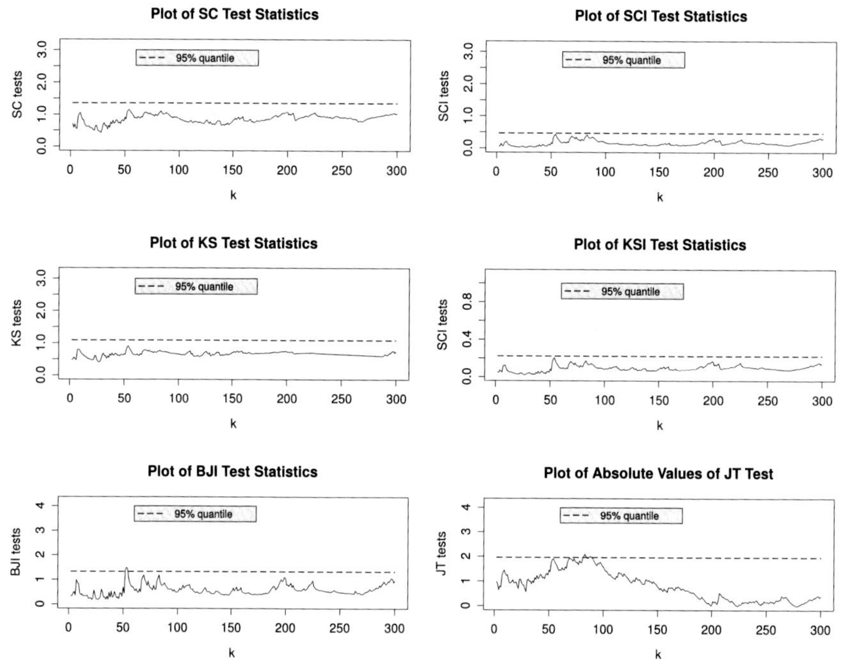


**Figure 5.6 Plots of goodness-of-fit tests for automobile bodily injury claims data.**

from a single state and can be found online from the accompanying data site in Frees [40].

We are interested in the distribution function for the claimed loss. We first look at the plots given in Fig. 5.5 for a visual examination. At the first glance, in a quite wide range of values of the sample fraction $k$, the Hill and alt-Hill plots look flat, but we can find from the Hill plot that upward and downward trends change alternatively several times, even one occurs at a small sample fraction $k = 29$ or approximately at $\theta = 0.46$ from the alt-Hill plot. To narrow the range for feasible sample fractions we run the six goodness-of-fit tests introduced in Chapter 2. The values of the six test statistics are plotted in Fig. 5.6, where the dashed lines are the 95% quantiles (or 5% critical values) for the limiting distributions of the corresponding goodness-of-fit test statistics. The values of the sample fraction $k$ in the range from 2 to 50 pass all six tests at 5% significance level. Since 29 falls within this range, we decide to take $k = 29$ as the turning point. The Hill plot is given in Fig. 5.7 for the range of $2 \leq k \leq 100$.

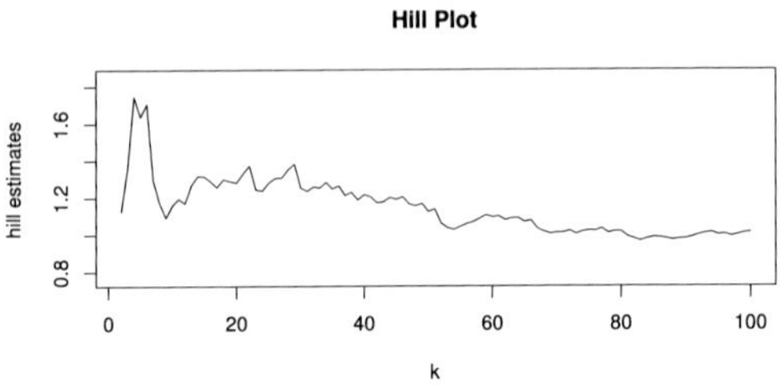


Figure 5.7 **Hill plot for automobile bodily injury claims** in a shorter range of $k$.

Table 5.1 Estimates of high quantiles for automobile bodily injury claims.

| Percentage | $p$ | Quantile estimate | Sample quantile |
|---|---|---|---|
| 99.0% | .010 | 69.93 | 66.89 |
| 99.5% | .005 | 115.49 | 114.60 |
| 99.8% | .002 | 224.13 | 193.00 |
| 99.9% | .001 | 370.11 | 222.41 |

**Estimating tail index.** After selecting $k = 29$, we estimate the tail index by the Hill estimate, which is $\hat{\alpha} = \hat{\alpha}(29) = 1.382$ with a standard deviation $\hat{\alpha}/\sqrt{k} = 0.2566$. Therefore, a 95% confidence interval for $\alpha$ is $1.382 \mp 1.96 \times 0.2566 = (0.879, 1.885)$ by treating the asymptotic bias of the Hill estimate with this selected sample fraction is zero.

**Estimating high quantiles.** We estimate 99% quantile for the distribution function of the claimed losses, which is denoted by $x_{0.01}$ as in Section 2.3 of Chapter 2. With the choice of $k = 29$, $X_{n,n-k} = 40$. By using the estimator $\hat{x}_p$ in (2.56) and Theorem 2.16, we have

$$\hat{x}_{0.01} = 40 \times (\frac{29}{1340 \times 0.01})^{1/1.382} = 69.93$$

with a standard deviation $69.93 \times \frac{\log(40/(1340\times 0.01))}{\sqrt{29}\times 1.382} = 7.255$. Hence a 95% confidence interval for $x_{0.01}$ is $69.93 \mp 1.96 \times 7.255 = (55.71, 84.15)$ by treating the asymptotic bias is zero.

In Table 5.1 we list some estimated high quantiles and the corresponding sample quantiles. We can see that the high quantile estimate at a higher level is significantly larger than the corresponding sample quantile, which is desirable from the risk management point of view.

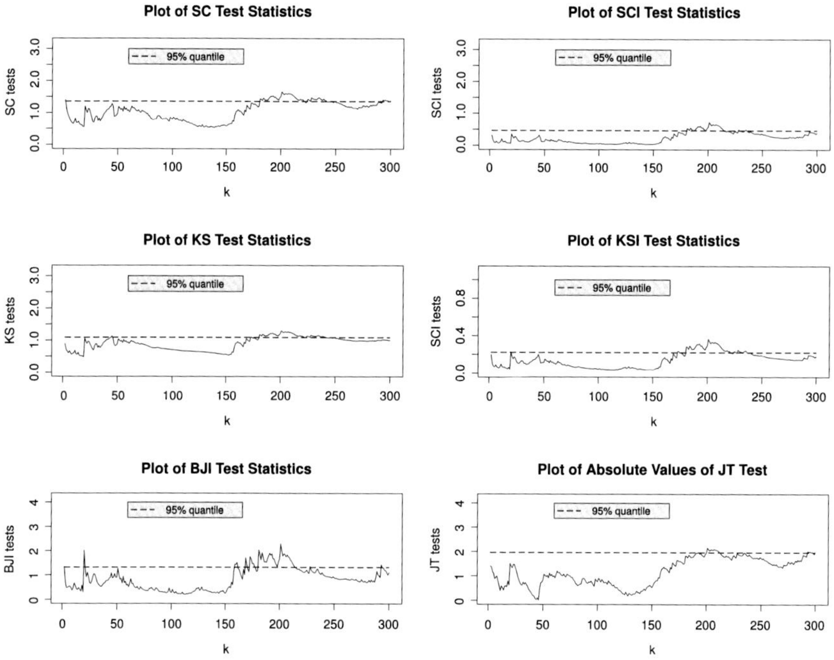


**Figure 5.8 Plots of goodness-of-fit tests for automobile insurance claims data.**

## 5.3.2 Automobile Insurance Claims

**Data description**. We consider an Automobile Insurance Claims dataset from Frees [40]. The dataset contains $n = 6773$ automobile insurance claims. The variable we are interested in is the amount paid on a closed claim, in US dollars from a large midwestern (US) property and casualty insurer for private passenger automobile insurance. The claims that were not closed by year end are handled separately. Insurers categorize policyholders according to a risk classification system. This insurer's risk classification system is based on automobile operator characteristics and vehicle characteristics which are also available in the dataset.

To assess the underlying distribution of the paid insurance claims, we run the six goodness-of-fit tests. The values of the six test statistics are plotted in Fig. 5.8. All six tests detect a heavy tail when the sample fraction $k$ is taken in the range from 2 to 155, except that BJI test fails only at $k = 20$. The Hill plots are given in Fig. 5.9, where a turning point appears at $k = 128$. Hence we will choose $k = 128$ and use it in our estimation below.

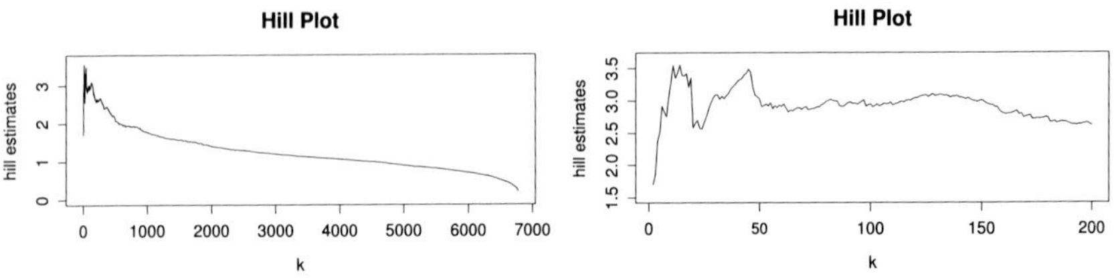


**Figure 5.9 Two Hill plots for automobile insurance claims data** in the full range and a short range of the sample fraction.

**Estimating tail index.** By setting $k = 128$, the Hill estimate gives $\hat{\alpha} = 3.104$, and its standard deviation is 0.272. Therefore, a 95% confidence interval for the tail index is (2.566, 3.642).

**Estimating high quantiles.** We can estimate high quantiles by using the estimator defined in (2.56). With $k = 128$, 99%, 99.5%, 99.8% and 99.9% quantiles are estimated to be 12,227.90, 15,287.25, 20,536.51, and 25,674.62, respectively. The corresponding sample quantiles are 12,037.05, 15,279.68, 21,404.96 and 24,624.44.

**Estimating mean.** From the analysis above we are confident that the distribution function for the automobile insurance claims is a heavy tailed distribution function with an index $\alpha > 2$. We can estimate the mean of the distribution by the sample mean which is 1853.04 with a standard error 32.16. We can also use the estimator for the mean in Section 2.7. The resulting estimate is 1852.47, which is really close to the sample mean. However the estimated standard deviation of this estimator is 24.70, which is smaller than 32.16. Theoretically both estimators have the same asymptotic variance. The observed difference may be due to the impact of a proportion of the largest observations used in estimating the standard deviation of the sample mean.

### 5.3.3 Hospital Costs

**Data description.** The Hospital Costs data were from the Nationwide Inpatient Sample of the Healthcare Cost and Utilization Project (NIS-HCUP), a nationwide survey of hospital costs conducted by the US Agency for Healthcare Research and Quality (AHRQ). The data we consider here are limited to Wisconsin hospitals and the dataset contains the hospital discharge costs in year 2003 from 500 claims for patients aged 0–17 years. See Frees [40] for more details on this dataset.

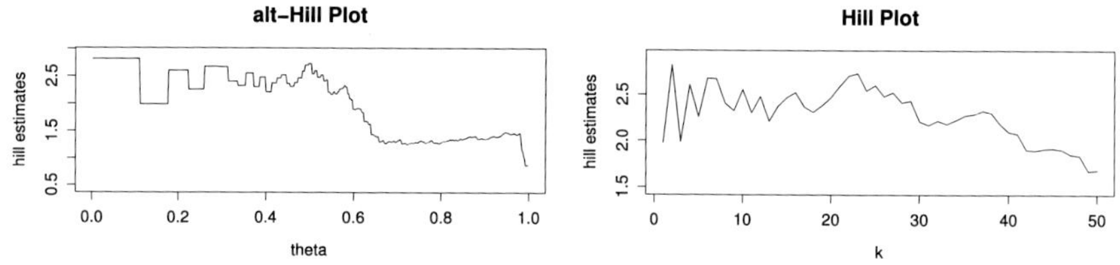


**Figure 5.10 alt-Hill plot and Hill plot** for Hospital costs data.

**Table 5.2** Estimates of the tail index and the mean for hospital costs.

| **Parameter** | **Estimate** | **Standard error** | 95% **confidence interval** |
|---|---|---|---|
| Tail index | 2.730 | 0.569 | (1.614, 3.845) |
| Mean | 2777.105 | 110.550 | (2560.43, 2993.78) |

We first examine the Hill plot and the alt-Hill plot. We note that both plots look stable in a small range of the sample fraction $k$. Fig. 5.10 contains the alt-Hill plot and the Hill plot in the range of $k$ from 1 to 50. All six goodness-of-fit tests do not reject the heavy tail hypothesis when the sample fraction $k$ is between 2 and 40. A turning point can be identified from the alt-Hill plot when $\theta$ is approximately 0.50 or from the Hill plot when $k = 23$. By using $k = 23$, the Hill estimate for the tail index is 2.703. We also estimate the mean of the distribution by using the method developed in Section 2.7. Table 5.2 gives both estimates for the tail index and the mean, and their 95% confidence intervals, where the estimate for the mean is very close to the sample mean 2774.388.

### 5.3.4 Danish Fire Losses Data

**Data description.** The data set consists of 2156 Danish fire losses over one million Danish krone from year 1980 to year 1990 inclusive, see Fig. 5.11. The loss is a total loss for the event concerned and includes damage to buildings, furnishings and personal property as well as loss of profits. This dataset has been embedded in **R** package "QRM" and it is also available at www.ma.hw.ac.uk/~mcneil/.

This Danish fire dataset has been analyzed by McNeil [73], Resnick [92] and Haug et al. [55], where the right tail index was confirmed to be between 1 and 2. Moreover Peng [80] constructed confidence intervals for the mean value of the Danish fire loss by using empirical likelihood methods.

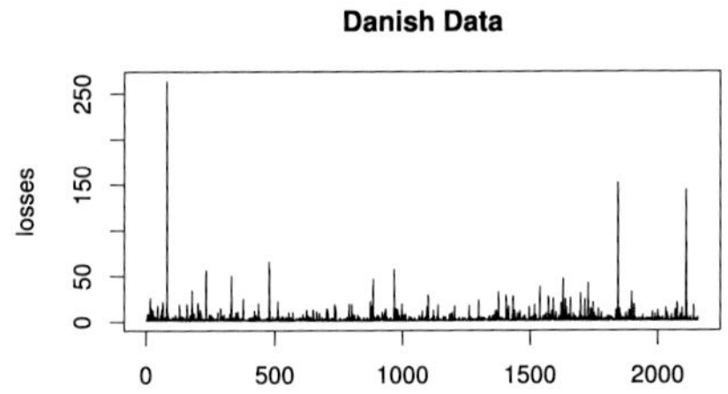


**Figure 5.11 Plot of Danish data** for fire losses over one million Danish krone from the years 1980 to 1990.

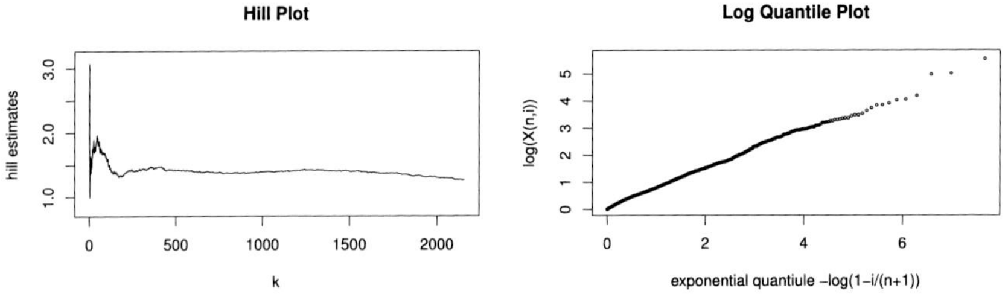


**Figure 5.12 Hill plot and log-quantile plot** of Danish data.

We provide the log quantile plot and the Hill plot in Fig. 5.12. From the log quantile plot we see a nearly straight line, and the Hill plot looks flat in a very wide range of sample fraction, say from 150 to 1500. Both plots may suggest the use of a large sample fraction which results in an estimate of the tail index between 1.4 and 1.5. But choosing an appropriate sample fraction may not be that easy for this example. We run the six goodness-of-fit tests and all these six tests accept the distribution with a heavy tail when the sample fraction is between 2 and 110. We further observe an upward trend followed by a downward trend roughly in the range $2 < k < 150$, and the Hill estimates fluctuate in this range of sample fraction. After we examine a shorter Hill plot in Fig. 5.13, we can find a turning point at $k = 46$ and an upward trend is obvious before this sample fraction. Even if we ignore the first few Hill estimates, the difference between the largest and smallest Hill estimates in this region is more than 0.6. And the Hill estimate at $k = 46$ is 1.969.

Next we try a different approach to identify $k$ by shifting the location of the data. We consider a location shift by subtracting 5 from each observation. For illustration, we use $x_1, \cdots, x_n$ to denote the original sample, and define the new sample as $y_i = x_i - 5$ for $1 \le i \le n$. Keep in mind that this

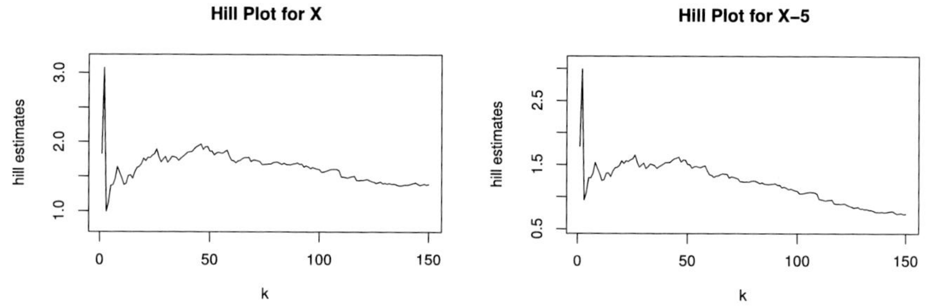


**Figure 5.13 Hill plots for Danish data** for original observations ($X$) and transformed ones ($X - 5$).

**Table 5.3** Comparison of estimates from the original data and shifted data.

| Data | Fraction $k$ | Estimate | Standard error | 95% confidence interval |
|---|---|---|---|---|
| Original | 46 | 1.969 | 0.2903 | (1.400, 2.538) |
| New | 46 | 1.617 | 0.2384 | (1.150, 2.084) |

will result in some negative observations in the new data set. The underlying distribution for the new data is different from that for the original data but they have the same tail index.

**Estimating tail index.** We can analyze the new data $y_1, \cdots, y_n$ using the same procedure as before. The Hill plot for transformed data is given in Fig. 5.13 for the range $1 \leq k \leq 150$. After we run the six goodness-of-fit tests we find out that all six tests detect a heavy tail in the range $2 \leq k \leq 60$. Then we search the turning point in this range. The turning point is identified as $k = 46$, which happens to be the same as the one for the original data. Comparing the two Hill plots in Fig. 5.13, the Hill estimates based on the shifted data are a bit more stable.

Table 5.3 reports point estimates of the tail index and their corresponding confidence intervals based on these two approaches. An assessment of the difference will be given later.

In Resnick [92], the tail index $\alpha$ for the Danish fire losses is estimated as $\hat{\alpha} = 1.4$ from smoothed Hill's estimates, and about 1500 largest observations are used in the estimation. It seems that too many observations have been used in the estimation. This can be seen from Fig. 5.14 as all six tests reject a heavy tail in a quite wide range of sample fraction $k$ including $k = 150$.

Next we construct confidence intervals by applying the empirical likelihood method introduced in Section 2.5.1.3 to the shifted data. From Table A.1, a 5%-level critical value for the empirical likelihood ratio statistic with $k = 46$ is 4.381. Hence a 95% confidence interval for the tail

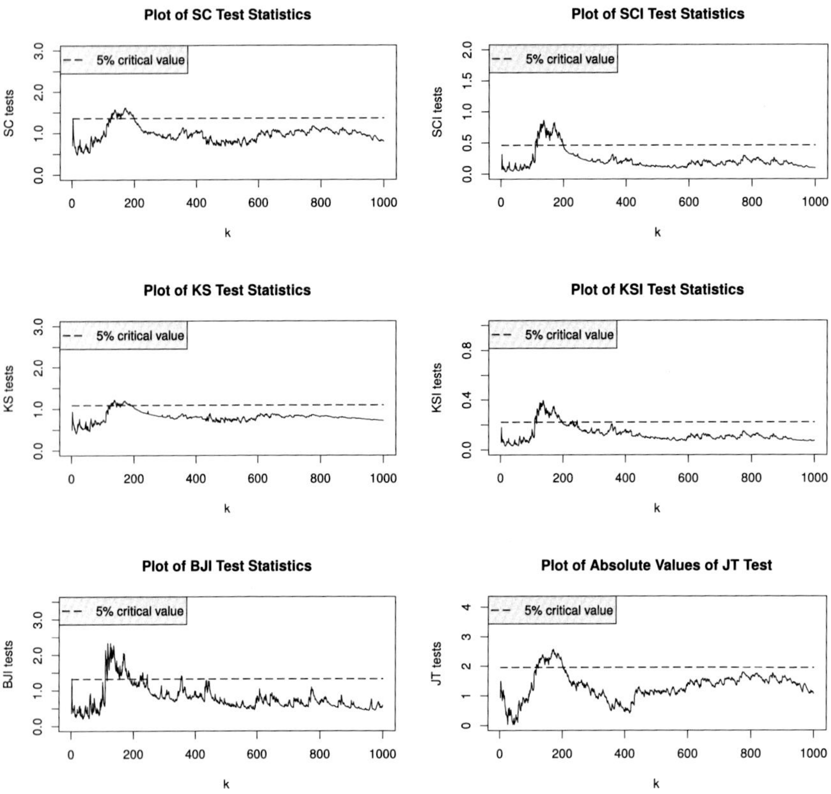


**Figure 5.14 Plots of tests for Danish loss data.**

index by applying the empirical likelihood method to the shifted data is obtained as $(1.227, 2.239)$, which is more skewed to the right than the normal approximation based confidence interval in Table 5.3. This is one of advantages of empirical likelihood based intervals/regions, where the shape of the obtained interval/region is automatically determined by the data.

**Assessing the estimates of the tail index.** Assume $\{x_1, \cdots, x_n\}$ is a random sample from the distribution function $F$. It is well known that a nonparametric estimate of the underlying distribution $F$ is given by

$$F_n(x) = \frac{1}{n}\sum_{i=1}^{n} I(x_i \le x), \qquad x \in \mathbb{R}.$$

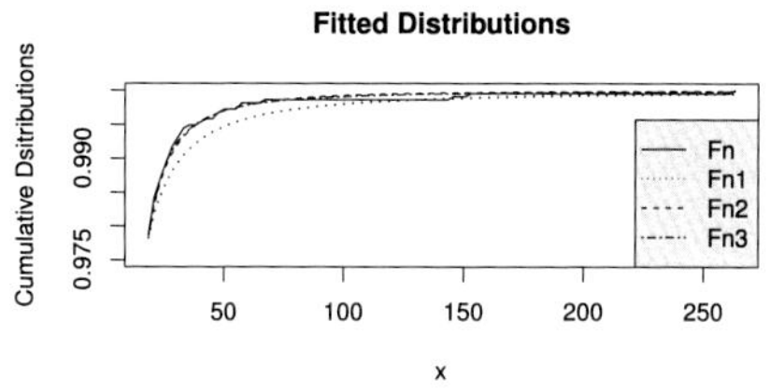


**Figure 5.15 Comparison of estimates for tail index** plot of empirical and estimated distributions for Danish fire losses.

To evaluate how well the estimated tail indices fit the data, we use Eq. (2.1) to estimate the underlying distribution function $F$. Let $a$ be an estimate of the tail index $\alpha$ and $T > 0$ be a high threshold. An estimate of $F(x)$, say $\hat{F}(x)$, can be defined as $F_n(x)$ for $x \leq T$ and

$$\hat{F}(x) = 1 - (1 - F_n(T))\left(\frac{x}{T}\right)^{-a} \quad \text{for} \quad x > T.$$

Now we choose $k = 46$, $T = X_{n,n-k} = X_{2156,2110} = 18.32208$. Based on Resnick's estimate $\hat{\alpha} = 1.4$, we can estimate $F(x)$ by

$$F_{n1}(x) = 1 - (1 - \frac{2109}{2156})\left(\frac{x}{18.32208}\right)^{-1.4} \quad \text{for} \quad x > 18.32208.$$

Based on our estimate $\hat{\alpha} = 1.969$ from the original data we estimate $F$ by

$$F_{n2}(x) = 1 - (1 - \frac{2109}{2156})\left(\frac{x}{18.32208}\right)^{-1.969} \quad \text{for} \quad x > 18.32208.$$

Finally, from the estimate $\hat{\alpha} = 1.612$ based on the shifted data we obtain the following estimate for the distribution $F$ as

$$F_{n3}(x) = 1 - (1 - \frac{2109}{2156})\left(\frac{x-5}{18.32208-5}\right)^{-1.617} \quad \text{for} \quad x > 18.32208.$$

We plot the four distribution functions $F_n(x)$, $F_{n1}(x)$, $F_{n2}(x)$ and $F_{n3}(x)$ for $x > 18.32208$ in Fig. 5.15, and conclude that both $F_{n2}$ and $F_{n3}$ fit the empirical distribution $F_n$ equally well in the upper tail, and the estimated distribution $F_{n1}$ based on the smoothed Hill estimate is slightly away from the empirical distribution in many data points. We notice that the Hill estimate in $F_{n1}$ uses a different $k$ from the corresponding $k$ in the employed threshold, which may affect the comparison.

**Estimating high quantiles.** We choose to estimate a high quantile under location shift. One can estimate a high quantile from $y_1, \cdots, y_n$ first and then add 5 to get a high quantile estimate for the underlying distribution. Assume we want to estimate a 99.9% high quantile $x_{0.001}$. Take $k = 46$ and $\hat{\alpha} = 1.617$. Then we estimate $y_{0.001} = x_{0.001} - 5$ by

$$\hat{y}_{0.001} = (18.32208 - 5) \times (\frac{46}{2156 \times 0.001})^{1/1.617} = 88.403,$$

and thus a 99.9% quantile is estimated by $\hat{x}_{0.005} = 88.403 + 5 = 93.403$. A 95% confidence interval for $x_{0.001}$ is (45.056, 141.749).

**Estimating mean.** Again we estimate the mean based on the shifted data $y_1, \cdots, y_n$ and add 5 to get an estimate of the mean for the original loss. We then obtain an estimate of the mean Danish fire loss over one million Danish Krone with 3.415492 and its 95% confidence interval is (3.085767, 3.745216). Note that the sample mean of the 2167 observations is 3.397257.

## 5.4 APPLICATIONS TO DEPENDENT DATA

An autoregressive integrated moving average (ARIMA) model is an extension of an autoregressive moving average (ARMA) model. For a time series $\{y_n\}$, define a first-order differencing operator by $\nabla y_n = y_n - y_{n-1}$. The $d$-th order differencing is defined recursively by $\nabla^d y_n = \nabla(\nabla^{d-1} y_n)$ for any $d \geq 1$ with convention $\nabla^1 y_n = \nabla y_n$ and $\nabla^0 y_n = y_n$.

Let $\{Y_n\}$ be a time series and $p, d, q \geq 0$ be integers. If $X_n = \nabla^d Y_n$ is a stationary sequence of random variables satisfying ARMA model (3.2), then $\{Y_n\}$ is said to follow an ARIMA($p, d, q$) model. If $d = 0$, then an ARIMA($p, 0, q$) model is an ARMA($p, q$) model.

The time series $\{Y_n\}$ generated from an ARIMA($p, d, q$) model may not be stationary. By applying differencing technique, we know that $X_n = \nabla^d Y_n$ is stationary and follows an ARMA($p, q$) model. Our aim is to confirm whether the distributions are heavy-tailed for the data in finance, insurance and some other areas. It follows from Theorem 3.5 and Remark 3.1 that the Hill estimator is still asymptotically normal under suitable conditions, which indicates that the Hill plots are still powerful tools for analyzing the time series data.

The limiting distributions for goodness-of-fit tests haven't been obtained for general dependent data, but the convergence of the six goodness-of-fit tests can be proved under certain technical conditions via the convergence of the tail empirical process and the tail quantile process under dependence in Chapter 3. Since the limiting distributions of tests depend on the dependence structure of the sequence, their quantiles or critical values are difficult to obtain in general. We have demonstrated in Chapter 3 that the tail index for a time series under certain ARMA models is the same as that for the errors, see, e.g., Eq. (3.6). The errors are not observable in practice and so our conducted inference for real data sets in this chapter will be based on the model residuals. When the coefficients in an ARMA model are well estimated, the model residuals and the true model errors are very close, and their differences are negligible compared with a few largest residuals on which our inference is based. Therefore, the inference procedures for the tail index developed for i.i.d. data are applicable to a time series model via the model residuals, see Theorem 3.5 and Remark 3.1. The goodness-of-fit tests and inference procedures for high quantiles developed for i.i.d. data are also expected to be applicable.

In our analysis below, we fit ARIMA$(p, d, q)$ models by using some **R** packages and the order parameters $p, d$ and $q$ are selected based on the Akaike information criterion (AIC), the corrected Akaike information criterion (AICc) and the Bayesian information criterion (BIC). For these criteria, we refer to Akaike [2], Sugiura [99], Hurvich and Tsai [59], and Schwarz [96].

Analysis of residuals is an import step to check the model assumptions in regression analysis including times series models. Some commonly used plots include (a) Residuals plot, (b) Normal QQ (quantile–quantile) plot, (c) ACF (sample autocorrelation function) plot, and (d) PACF (sample partial autocorrelation function) plot.

### 5.4.1 Daily Foreign Exchange Rates

**Data description.** The dataset contains daily foreign exchange rates between eight foreign currencies and US dollars from December 31, 1979 to December 31, 1998. Eight variables, Australia/US, British/US, Canadian/US, Dutch/US, French/US, German/US, Japanese/US, and Swiss/US, are daily exchange rates in units of foreign current per US dollar. The data set is available from *Time Series Data Library* (http://datamarket.com/data/list/?q=provider:tsdl).

We analyze three exchange rates, French/US, German/US, and Swiss/US. We use ARIMA models to fit these data and identify whether the distributions of the errors are heavy-tailed. We employ the program "auto.arima" in **R** package "forecast" to select the best orders for fitting

**Table 5.4** Model information on exchange rates.

| Currency | Model | Tested range | Selected $k$ | Estimate of index |
|---|---|---|---|---|
| French/US | ARIMA(2, 1, 2) | [10, 310] | 79 | 4.260 |
| German/US | ARIMA(0, 1, 1) | [2, 55] | 45 | 5.363 |
| Swiss/US | ARIMA(1, 1, 0) | [2, 130] | 104 | 5.012 |

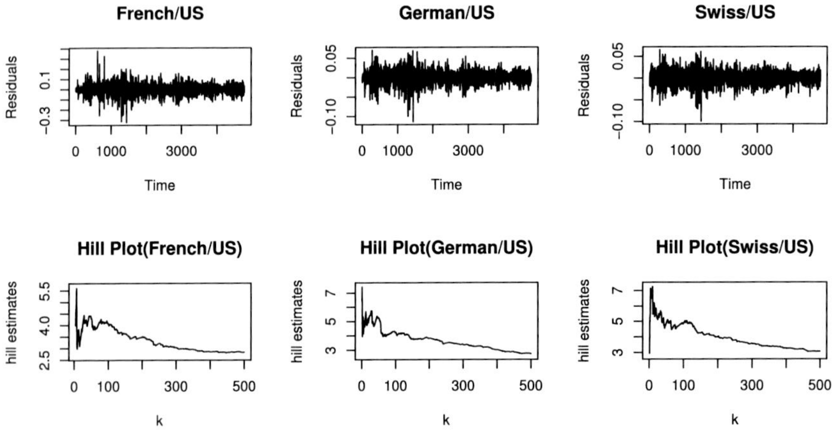


**Figure 5.16 Residuals plots and Hill plots** for exchanges rates of French/US, German/US and Swiss/US from December 31, 1979 to December 31, 1998.

an ARIMA model to each variable. Table 5.4 contains results on model selections including orders of the selected ARIMA models, the ranges of the sample fraction allowed by the goodness-of-fit tests, selected sample fractions and the Hill estimates for the tail indices. The residuals plots and the Hill plots are given in Fig. 5.16.

A common feature for these three time series of exchange rates is that the first differencing, $X_n = Y_n - Y_{n-1}$ follows an ARMA model. The estimates of the coefficients in the ARMA models are given as follows:

- French/US: $\hat{\phi}_1 = -0.0401$, $\hat{\phi}_2 = 0.6930$, $\hat{\theta}_1 = 0.0851$, $\hat{\theta}_2 = -0.6953$;
- German/US: $\hat{\theta}_1 = 0.0357$;
- Swiss/US: $\hat{\phi}_1 = 0.0386$.

Confidence intervals for the tail indices based on the Hill estimates for the above models can also be obtained. From Section 3.2 the Hill estimates based on residuals behave as if the residuals were independent. For Swiss/US exchange rates, a 95% confidence interval for the tail index is $5.012 \mp 1.96 \times 5.012/\sqrt{104} = (4.049, 5.976)$.

Some further discussions on this example are given in Section 5.5.

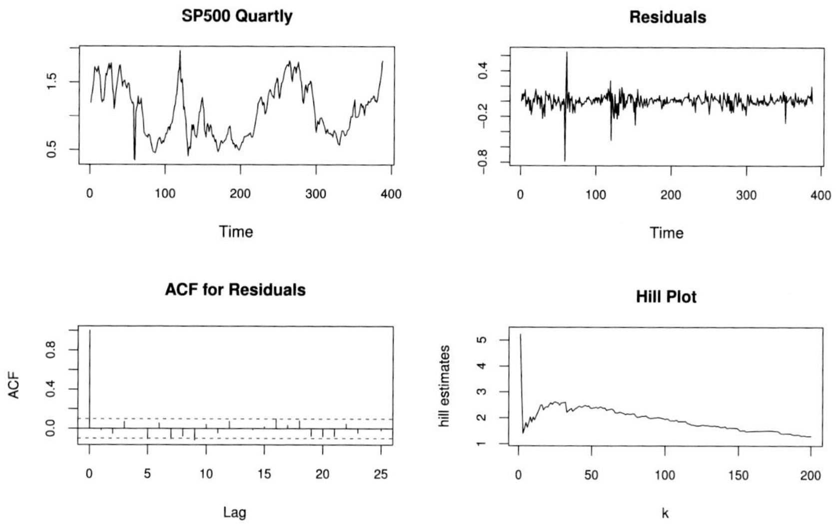


**Figure 5.17 Plots for SP500 quarterly returns** data plot, residuals plot, ACF plot, and Hill plot under ARIMA(0, 1, 1) model.

## 5.4.2 Quarterly S&P 500 Indices

**Data description.** The dataset contains a time series of Quarterly S&P 500 index from Feb 15, 1900 to Nov 15, 1996 with a total of 388 data points. This dataset is available online https://datamarket.com/data/set/22rk/quarterly-sp-500-index-1900-1996#s1493794198151 from "Time Series Data Library" in Hyndman [60].

We fit an ARIMA(0, 1, 1) model, that is, after the first differencing, the sequence satisfies an MA(1) model. The MA coefficient for the model is estimated to be $\hat{\theta}_1 = 0.2597$.

The original data, the residuals from the selected model, the ACF function of the residuals and the Hill plot are given in Fig. 5.17.

A common range for the sample fraction $k$ to pass all six goodness-of-fit tests is $2 \leq k \leq 32$. After a closer look at the Hill plot we choose $k = 32$ and the Hill estimate with this sample fraction is 2.607.

## 5.4.3 S&P 500 Weighted Daily Returns

**Data description.** The dataset consists of 1759 daily returns of the Standard and Poor's (S&P) value weighted index from year 2000 to year 2006. Each calendar year, there are about 250 days on which the exchange is open and stocks are traded. There are several indices to measure the mar-

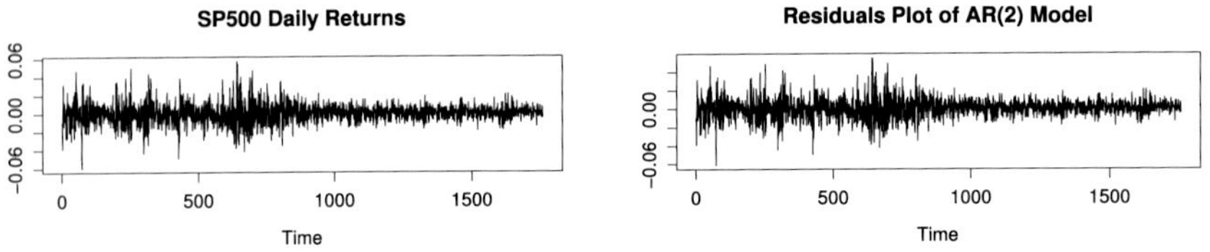


**Figure 5.18 Plots for SP500 weighted daily returns** data plot and residuals plot under AR(2) model.

ket's overall performance. The value weighted index is created by assuming that the amount invested in each stock is proportional to its market capitalization which is defined as the beginning price per share times the number of outstanding shares. The S&P equally weighted index for each trading day is an average of the closing price of various stocks on that day. The plot of the data is given in Fig. 5.18, and the data set is taken from Frees [40].

**Model building.** We first try to fit an ARIMA model. The model selection from auot.arima chooses the best model ARIMA(2, 0, 0), i.e. AR(2) model. We then fit an AR(2) model and plot the residuals in Fig. 5.18. The residuals plot clearly indicates that the volatility decays with time. This suggests a GARCH error for the AR(2). Therefore, we will build AR(2)–GARCH(1, 1) model:

$$x_n = \mu + \phi_1 x_{n-1} + \phi_2 x_{n-2} + \varepsilon_n, \quad \varepsilon_n = h_n \eta_n, \quad h_n^2 = \omega + a\varepsilon_{n-1}^2 + bh_{n-1}^2 \tag{5.1}$$

where $\{\eta_n\}$ is a sequence of independent and identically distributed random variables with mean zero and variance one.

We obtain the following estimates for the parameters in model (5.1):

$$\hat{\mu} = 0.0005213, \ \hat{\phi}_1 = -0.04102, \ \hat{\phi}_2 = -0.06236,$$
$$\hat{\omega} = 0.00000055, \ \hat{a} = 0.06626, \ \hat{b} = 0.9296,$$

and their standard errors are 0.000196, 0.02465, 0.02456, 0.0000002.687, 0.01082 and 0.01112, respectively.

Next we examine the residuals under AR(2)–GARCH(1, 1) model. The standardized residuals, which are defined as the residuals of the model divided by the estimates of the conditional standard deviations $h_n$'s, are estimates of $\eta_n$'s. The residuals plot in Fig. 5.19 seems quite stable, indicating a constant variance over the time. From both the ACF plot and the PACF plot, the sample autocorrelation function and the sample partial autocorrelation function cut off at all lags.

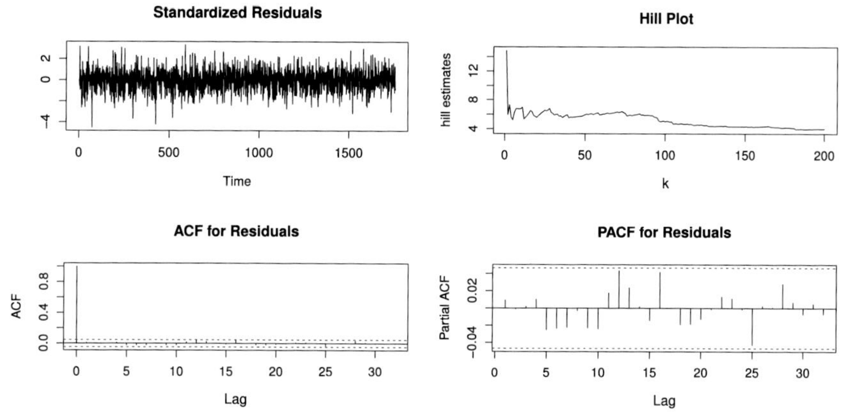


**Figure 5.19 Plots for SP500 weighted daily returns** residuals plot, Hill plot, ACF and PACF plots for residuals plots under AR(2)–GARCH(1, 1) model.

**Estimating tail index.** All six goodness-of-fit tests detect a heavy tail at level 0.05 for a common range of $k$ between 2 and 95. We pick up a value of $k = 73$ as a turning point. The Hill estimate for the tail index at this sample fraction is 6.373, and the corresponding 95% confidence interval for the tail index is $(4.911, 7.834)$.

**One-step-ahead prediction.** Under model (5.1), the predictor for $x_{n+1}$ based on $x_j$, $1 \le j \le n$ is called a one-step-ahead predictor, denoted by $x_{n:n+1}$, and it is given by

$$x_{n:n+1} = \mu + \phi_1 x_n + \phi_2 x_{n-1}.$$

By replacing the unknown parameters in the above equation by their estimates, we estimate $x_{n:n+1}$ by

$$\hat{x}_{n:n+1} = \hat{\mu} + \hat{\phi}_1 x_n + \hat{\phi}_2 x_{n-1}.$$

It is easy to see that $x_{n+1} - \hat{x}_{n:n+1}$ is approximately equal to $\varepsilon_{n+1} = h_{n+1}\eta_{n+1}$. Here $h_{n+1}$ can be estimated from (5.1) as

$$\hat{h}_{n+1}^2 = \hat{\omega} + \hat{a}\hat{\varepsilon}_n^2 + \hat{b}\hat{h}_n^2.$$

Therefore, the distribution of $(x_{n+1} - \hat{x}_{n:n+1})/\hat{h}_{n+1}$ can be well approximated by that of $\eta_{n+1}$. When the distribution of $\eta_{n+1}$ is symmetric, an approximate $(1-p)$ prediction interval for $x_{n+1}$ is $(\hat{x}_{n:n+1} - \hat{h}_{n+1}\bar{x}_{p/2}, \hat{x}_{n:n+1} + \hat{h}_{n+1}\bar{x}_{p/2})$, where $\bar{x}_{p/2}$ is $1-p/2$ quantile for the distribution of $|\eta_{n+1}|$. If the distribution of $\eta_{n+1}$ is not symmetric, we can use $p/2$ and $1-p/2$ quantiles for the distribution of $\eta_{n+1}$. If $p$ is not too small, we can simply use the $(1-p/2)$ sample quantile based on the standardized residuals $|\hat{\eta}_1|, \cdots, |\hat{\eta}_n|$. For a smaller $p$, we use the high quantile estimator introduced in Section 2.3. For this example, from the residuals plot we can see the residuals exhibit a good symmetry.

The sample size for this example is $n = 1759$. Since $x_n = -0.0014479$ and $x_{n-1} = -0.0044554$, we can easily get a one-step-ahead prediction

$$\begin{aligned}\hat{x}_{n:n+1} &= 0.0005213 - 0.04102 \times (-0.0014479) - 0.06236 \times 0.0044554 \\ &= 0.0007944.\end{aligned}$$

Suppose we want to estimate a 98% prediction confidence interval of $x_{n+1}$. Then $p/2 = 0.01$. The 99% quantile estimate given in Section 2.3 is 2.672772 when $k = 73$. The estimated conditional volatilities $h_j$'s for $j \le n$ and the standardized residuals $\hat{\eta}_j$'s, $j \le n$ can be obtained from the "fGARCH" package. The estimated $h_{n+1}$ is $\hat{h}_{n+1} = 0.005101$. Hence a 98% prediction interval for the future 1760-th observation is $(-0.01284, 0.01443)$.

## 5.5 SOME COMMENTS

In the previous two sections we have analyzed some independent datasets and dependent datasets in insurance and finance, and have seen evidence of heavy-tail distributions. In fact, heavy-tail phenomena may exist in many fields in practice.

Modeling a distribution is very important in statistics and can be challenging sometimes. For inference of some quantities such as mean and variance of a distribution and coefficients in classic regression models and time series models, consistency and convergence in distribution of estimators for these quantities usually require only certain conditions on some moments of the underlying distribution function. Even though, for inference such as constructing prediction confidence intervals under these models, a parametric distribution has to be specified or at least the underlying distribution should be well estimated to a certain degree. The extreme value theory can play an important role when some of these conditions fail. We have estimated the tail index and the mean from a heavy-tailed distribution and constructed prediction intervals in ARIMA–GARCH models

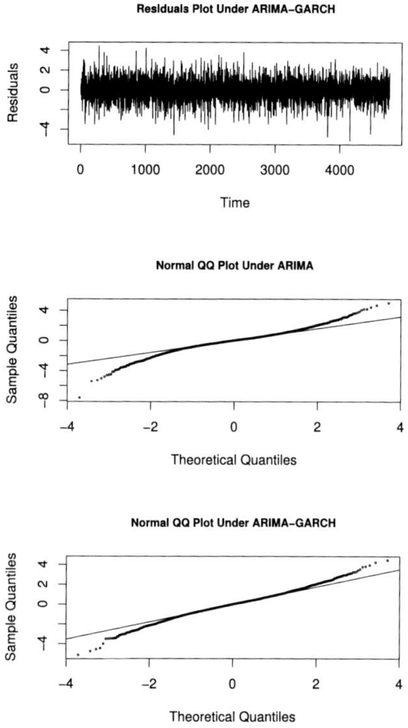


**Figure 5.20 Plots for exchange rates Sweiss/US** from ARIMA model to ARIMA–GARCH model.

by using methodologies developed in extreme value theory, while classical methods are not effective due to either the lack of moments or the interest in the tail region.

Every statistical method is based on certain model assumptions. When we pick up the sample fraction for the Hill estimator, carrying out some goodness-of-fit tests is highly recommended. Similarly, as we have assumed that the model errors in ARIMA/GARCH models are independent and identically distributed, analysis on model residuals such as residuals plots, ACF plots and PACF plots is recommended to ensure the validity of the employed model. Let us reexamine the example in Section 5.4.1 where three exchange rates are studied. If we check the residuals plots, more or less the heteroscedasticity exists. Based on the fitted ARIMA(1, 1, 0) model in Section 5.4.1, an improvement can be achieved in this case by fitting an ARIMA(1, 1, 0)–GARCH(1, 1) model. Three plots are given in Fig. 5.20. The first plot is the new residuals plot which has improved much upon the residuals under ARIMA(1, 1, 0) model in terms of homogeneity assumption. The second and the third plots are normal QQ-plots. For comparison, the residuals under ARIMA(1, 1, 0) model have been standardized by its sample standard deviation. The two straight lines in the plots pass through the origin (0, 0) with a slop 1. The two plots indicate a heavy tail but the tails in the normal QQ-plot under ARIMA(1, 1, 0)–GARCH(1, 1) do not look as wildly as those in the normal QQ-plot under ARIMA(1, 1, 0) model. In this case, an ARIMA(1, 1, 0)–GARCH(1, 1) may be preferred.

# APPENDIX A

# Tables

**Table A.1** Table for critical values of empirical likelihood for tail index.

| $k\backslash\alpha$ | **0.10** | **0.05** | **0.01** | $k\backslash\alpha$ | **0.10** | **0.05** | **0.01** |
|---|---|---|---|---|---|---|---|
| 10 | 4.735 | 8.189 | $\infty$ | 41 | 3.072 | 4.471 | 8.362 |
| 11 | 4.473 | 7.505 | 22.64 | 42 | 3.058 | 4.452 | 8.242 |
| 12 | 4.287 | 7.025 | 19.66 | 43 | 3.050 | 4.436 | 8.183 |
| 13 | 4.150 | 6.678 | 17.90 | 44 | 3.039 | 4.410 | 8.178 |
| 14 | 4.019 | 6.369 | 16.16 | 45 | 3.027 | 4.394 | 8.118 |
| 15 | 3.908 | 6.153 | 15.13 | 46 | 3.020 | 4.381 | 8.062 |
| 16 | 3.819 | 5.944 | 14.05 | 47 | 3.013 | 4.372 | 8.038 |
| 17 | 3.748 | 5.802 | 13.45 | 48 | 3.004 | 4.354 | 7.992 |
| 18 | 3.669 | 5.645 | 12.82 | 49 | 2.997 | 4.346 | 7.952 |
| 19 | 3.601 | 5.510 | 12.14 | 50 | 2.996 | 4.339 | 7.930 |
| 20 | 3.549 | 5.385 | 11.65 | 51 | 2.983 | 4.328 | 7.881 |
| 21 | 3.509 | 5.313 | 11.29 | 52 | 2.978 | 4.311 | 7.880 |
| 22 | 3.460 | 5.216 | 10.94 | 53 | 2.971 | 4.295 | 7.789 |
| 23 | 3.423 | 5.138 | 10.69 | 54 | 2.970 | 4.297 | 7.762 |
| 24 | 3.387 | 5.077 | 10.34 | 55 | 2.966 | 4.280 | 7.788 |
| 25 | 3.363 | 5.022 | 10.13 | 56 | 2.966 | 4.289 | 7.745 |
| 26 | 3.331 | 4.946 | 9.895 | 57 | 2.952 | 4.265 | 7.708 |
| 27 | 3.299 | 4.894 | 9.713 | 58 | 2.946 | 4.242 | 7.671 |
| 28 | 3.271 | 4.853 | 9.583 | 59 | 2.938 | 4.247 | 7.666 |
| 29 | 3.258 | 4.803 | 9.458 | 60 | 2.947 | 4.246 | 7.646 |
| 30 | 3.223 | 4.757 | 9.287 | 61 | 2.933 | 4.223 | 7.654 |
| 31 | 3.206 | 4.725 | 9.152 | 62 | 2.939 | 4.229 | 7.606 |
| 32 | 3.185 | 4.686 | 9.028 | 63 | 2.920 | 4.201 | 7.581 |
| 33 | 3.176 | 4.659 | 8.953 | 64 | 2.910 | 4.190 | 7.551 |
| 34 | 3.161 | 4.624 | 8.793 | 65 | 2.918 | 4.191 | 7.504 |
| 35 | 3.144 | 4.593 | 8.773 | 66 | 2.908 | 4.189 | 7.539 |
| 36 | 3.126 | 4.585 | 8.672 | 67 | 2.906 | 4.179 | 7.476 |
| 37 | 3.120 | 4.568 | 8.629 | 68 | 2.903 | 4.170 | 7.461 |
| 38 | 3.101 | 4.526 | 8.490 | 69 | 2.899 | 4.170 | 7.461 |
| 39 | 3.089 | 4.517 | 8.482 | 70 | 2.897 | 4.159 | 7.428 |
| 40 | 3.078 | 4.474 | 8.390 | 71 | 2.902 | 4.169 | 7.433 |

*continued on next page*

**Table A.1** *(continued)*

| $k\backslash\alpha$ | 0.10 | 0.05 | 0.01 | $k\backslash\alpha$ | 0.10 | 0.05 | 0.01 |
|---|---|---|---|---|---|---|---|
| 72 | 2.894 | 4.157 | 7.412 | 86 | 2.864 | 4.109 | 7.290 |
| 73 | 2.896 | 4.153 | 7.412 | 87 | 2.854 | 4.096 | 7.260 |
| 74 | 2.892 | 4.157 | 7.407 | 88 | 2.849 | 4.083 | 7.254 |
| 75 | 2.884 | 4.145 | 7.388 | 89 | 2.854 | 4.092 | 7.233 |
| 76 | 2.886 | 4.149 | 7.380 | 90 | 2.852 | 4.079 | 7.202 |
| 77 | 2.885 | 4.144 | 7.403 | 91 | 2.850 | 4.091 | 7.263 |
| 78 | 2.883 | 4.134 | 7.372 | 92 | 2.839 | 4.072 | 7.193 |
| 79 | 2.874 | 4.118 | 7.328 | 93 | 2.858 | 4.088 | 7.243 |
| 80 | 2.873 | 4.113 | 7.327 | 94 | 2.847 | 4.084 | 7.205 |
| 81 | 2.866 | 4.107 | 7.320 | 95 | 2.848 | 4.069 | 7.188 |
| 82 | 2.871 | 4.122 | 7.315 | 96 | 2.839 | 4.067 | 7.176 |
| 83 | 2.860 | 4.097 | 7.299 | 97 | 2.850 | 4.087 | 7.204 |
| 84 | 2.851 | 4.092 | 7.252 | 98 | 2.846 | 4.061 | 7.165 |
| 85 | 2.868 | 4.115 | 7.317 | 99 | 2.832 | 4.056 | 7.163 |

The $\alpha$-level critical values for $\alpha = 10\%$, 5% and 1% for all $k$ between 10 and 99 are generated from the empirical likelihood statistics with 1,000,000 random samples from the standard exponential distribution.

**Table A.2** Table for limiting critical values of goodness-of-fit tests.

| **Tests** | **$\alpha$ level critical values of tests** | | | | | | |
|---|---|---|---|---|---|---|---|
| | *0.15* | *0.10* | *0.05* | *0.02* | *0.01* | *0.005* | *0.001* |
| Estimated score (SC) | 1.138 | 1.224 | 1.358 | 1.517 | 1.628 | 1.731 | 1.949 |
| Integrated score (SCI) | 0.285 | 0.348 | 0.461 | 0.618 | 0.743 | 0.871 | 1.179 |
| Kolmogorov–Smirnov (KS) | 0.930 | 0.992 | 1.090 | 1.208 | 1.288 | 1.366 | 1.526 |
| Cramer–von Mises (KSI) | 0.148 | 0.174 | 0.222 | 0.286 | 0.338 | 0.388 | 0.514 |
| Integrated Berk–Jones (BJI) | 0.915 | 1.062 | 1.321 | 1.680 | 1.956 | 2.239 | 2.936 |

[a] The above critical values are based on the distributions of the random variables defined on the right-hand sides of Eqs. (2.63) and (2.64) in Chapter 2 as limits of the standardized test statistics defined on the left-hand sides of the two equations.

[b] The SC test statistic has a limiting distribution $G(x) = 1 - 2\sum_{i=1}^{\infty}(-1)^{i-1}e^{-i^2x^2}$ for $x > 0$. The limiting critical values for this test in the table is obtained from the inverse of the distribution $G$.

[c] The limiting critical values for SCI, KS, KSI and BJI tests are obtained by simulations. By using **R** package **e1071**, 500,000 random samples of Brownian motion on $[0, 1]$ with 50,000 equally spaced grid points are used to estimate the limiting critical values for these test statistics.

# APPENDIX B

# List of Notations and Abbreviations

| | |
|---|---|
| $\mathbb{R}$ | The set of all real numbers, $(-\infty, \infty)$ |
| $\mathbb{R}^d$ | $d$-dimensional Euclidean space |
| $\mathbb{R}_+$ | The set of all nonnegative numbers |
| $\bar{\mathbb{R}}$ | The closure of $\mathbb{R}$, $[-\infty, \infty]$ |
| $RV_\rho^a$ | The set of all regularly varying functions at $a$ with exponent $\rho$, where $a = 0$ or $\infty$ |
| $D(G)$ | The set of all distributions in the domain of attraction of an extreme-value distribution $G$ |
| $\mathbf{N}(\mu, \sigma^2)$ | Univariate normal distribution with mean $\mu$ and variance $\sigma^2$ |
| $\mathbf{N}(\mu, \Sigma)$ | Multivariate normal distribution with mean vector $\boldsymbol{\mu}$ and covariance matrix $\Sigma$ |
| $F^-$ | The generalized inverse of a non-decreasing function $F$ |
| $\log^+(\cdot)$ | $\log(\max(\cdot, 1))$ |
| $\xrightarrow{D}$ | Weak convergence |
| $\xrightarrow{v}$ | Vague convergence |
| $\xrightarrow{d}$ | Convergence in distribution |
| $\xrightarrow{p}$ | Convergence in probability |
| $\stackrel{d}{=}$ | Equality in distribution |
| $\mathcal{N}$ | The set of all positive integers |
| $\lvert\mathbf{x}\rvert_r$ | $L_r$-norm of d-dimensional vector $\mathbf{x} = (x_1, \cdots, x_d)^T$, $(\sum_{i=1}^d \lvert x_i \rvert^r)^{1/r}$ |
| $\mathbf{max(x, y)}$ | $(\max(x_1, y_1), \cdots, \max(x_d, y_d))^T$ for $\mathbf{x} = (x_1, \cdots, x_d)^T$, $\mathbf{y} = (y_1, \cdots, y_d)^T$ |
| $\mathbf{x/y}$ | $(x_1/y_1, \cdots, x_d/y_d)^T$ for $\mathbf{x} = (x_1, \cdots, x_d)^T$, $\mathbf{y} = (y_1, \cdots, y_d)^T$ |
| $\mathbf{xy}$ | $(x_1 y_1, \cdots, x_d y_d)^T$ for $\mathbf{x} = (x_1, \cdots, x_d)^T$, $\mathbf{y} = (y_1, \cdots, y_d)^T$ |
| $\det A$ | Determinant of a matrix $A$ |
| $I(B)$ | Indicator function of the set $B$ |
| **LADE** | Least absolute deviations estimator (or estimation) |
| **i.i.d.** | Independent and identically distributed |
| **QMLE** | Quasi-maximum likelihood estimator (or estimation) |
| **ARIMA** | Autoregressive integrated moving average model |
| **ARMA** | Autoregressive moving average model |
| **ACF** | Autocorrelation function |
| **PACF** | Partial autocorrelation function |
| **GARCH** | Generalized autoregressive conditionally heteroscedastic process |

# BIBLIOGRAPHY


[1] J. Ahn, N. Shyamalkumar, Asymptotic theory for the empirical Haezendonck–Goovaerts risk measure, Insurance: Mathematics and Economics 55 (2014) 78–90.
[2] H. Akaike, A new look at the statistical model identification, IEEE Transactions on Automatic Control 19 (1974) 716–723.
[3] J. Angus, Asymptotic theory for bootstrapping the extremes, Communications in Statistics – Theory and Methods 22 (1993) 15–30.
[4] B. Basrak, R.A. Davis, T. Mikosch, A characterization of multivariate regular variation, Annals of Applied Probability 12 (2002) 908–920.
[5] J. Beirlant, T. de Wet, Y. Goegebeur, Statistics of Extremes: Theory and Applications, John Wiley & Sons Ltd, New Jersey, 2004.
[6] J. Beirlant, T. de Wet, Y. Goegebeur, A goodness-of-fit statistic for Pareto-type behaviour, Journal of Computational and Applied Mathematics 186 (2006) 99–116.
[7] I. Berkes, L. Horváth, Limit results of the empirical process of squared residuals in GARCH models, Stochastic Processes and Their Applications 105 (2003) 271–298.
[8] I. Berkes, L. Horváth, P. Kokoszka, Estimation of the maximal moment exponent of a GARCH(1, 1) sequence, Econometric Theory 19 (2003) 565–586.
[9] N. Bingham, C. Goldie, J. Teugels, Regular Variation, Cambridge University Press, Cambridge, 1987.
[10] M. Borkovec, C. Klüppelberg, The tail of the stationary distribution of an autoregressive process with ARCH(1) errors, Annals of Applied Probability 11 (2001) 1220–1241.
[11] F. Caeiro, M. Gomes, A new class of estimators of a "scale" second order parameter, Extremes 9 (2006) 193–211.
[12] J. Cai, J. Einmahl, L. de Haan, Estimation of extreme risk regions under multivariate regular variation, Annals of Statistics 39 (2011) 1803–1826.
[13] N. Chan, L. Peng, Weighted least absolute deviations estimation for an AR(1) process with ARCH(1) errors, Biometrika 92 (2005) 477–484.
[14] N. Chan, L. Tran, On the first-order autoregressive process with infinite variance, Econometric Theory 5 (1989) 354–362.
[15] N. Chan, R. Zhang, Inference for unit-root models with infinite variance GARCH errors, Statistica Sinica 20 (2010) 1363–1393.
[16] N. Chan, S. Deng, L. Peng, Z. Xia, Interval estimation for the conditional value-at-risk based on GARCH models with heavy tailed innovations, Journal of Econometrics 137 (2007) 556–576.
[17] N. Chan, L. Peng, R. Zhang, Interval estimation for GARCH(1, 1) tail index, Test 21 (2012) 546–565.
[18] N. Chan, D. Li, L. Peng, R. Zhang, Interval estimation of the tail index of an AR(1) with ARCH(1) errors, Econometric Theory 29 (2013) 920–940.
[19] S. Cheng, J. Pan, Asymptotic expansions of estimators for the tail index with applications, Scandinavian Journal of Statistics 25 (1998) 717–728.
[20] S. Cheng, L. Peng, Confidence intervals for the tail index, Bernoulli 7 (2001) 751–760.
[21] M. Cheng, L. Peng, Regression modeling for nonparametric estimation of distribution and quantile functions, Statistics Sinica 12 (2002) 1043–1060.

[22] S. Coles, An Introduction to Statistical Modeling of Extreme Values, Springer, New York, 2001.
[23] S. Csörgő, L. Viharos, Asymptotic normality of least-squares estimators of tail indices, Bernoulli 3 (1997) 351–370.
[24] S. Csörgő, P. Deheuvels, D. Mason, Kernel estimates of the tail index of a distribution, Annals of Statistics 13 (1985) 1050–1077.
[25] M. Csörgő, S. Csörgő, L. Horváth, D. Mason, Weighted empirical and quantile processes, Annals of Probability 14 (1986) 31–85.
[26] J. Danielsson, L. de Haan, L. Peng, C.G. de Vries, Using a bootstrap method to choose the sample fraction in tail index estimation, Journal of Multivariate Analysis 76 (2001) 226–248.
[27] L. de Haan, A. Ferreira, Extreme Value Theory: An Introduction, Springer, New York, 2006.
[28] L. de Haan, U. Stadtmüller, Generalized regular variation of second order, Journal of Australian Mathematical Society (Series A) 61 (1996) 381–395.
[29] A. Demattes, S. Clémencon, On tail index estimation based on multivariate data, Journal of Nonparametric Statistics 28 (2016) 152–176.
[30] G. Draisma, L. de Haan, L. Peng, T.T. Pereira, A bootstrap-based method to achieve optimality in estimating the extreme-value index, Extremes 2 (1999) 367–404.
[31] H. Drees, Weighted approximations of tail processes for $\beta$-mixing random variables, Annals of Applied Probability 10 (2000) 1274–1301.
[32] H. Drees, Extreme quantile estimation for dependent data, with applications to finance, Bernoulli 9 (2003) 617–657.
[33] H. Drees, E. Kaufmann, Selecting the optimal sample fraction in univariate extreme value estimation, Stochastic Processes and their Applications 75 (1998) 149–172.
[34] H. Drees, L. de Haan, S. Resnick, How to make a Hill plot, Annals of Statistics 28 (2000) 254–274.
[35] C. El-Nouty, A. Guillou, On the bootstrap accuracy of the Pareto index, Statistical Decision 18 (2000) 275–290.
[36] P. Embrechts, C. Klüppelberg, T. Mikosch, Modelling Extremal Events for Insurance and Finance, Springer, New York, 1997.
[37] W. Feller, An Introduction to Probability Theory and Its Applications, Volume II, second ed., Wiley, New York, 1971.
[38] A. Feuerverger, P. Hall, Estimating a tail exponent by modelling departure from a Pareto distribution, Annals of Statistics 27 (1999) 760–781.
[39] C. Francq, J. Zakoïan, Maximum likelihood estimation of pure GARCH and ARMA-GARCH processes, Bernoulli 10 (2004) 605–663.
[40] E.W. Frees, Regression Modeling with Actuarial and Financial Applications, Cambridge University Press, Cambridge, 2010. Datasets are available http://instruction.bus.wisc.edu/jfrees/jfreesbooks/default.aspx.
[41] S. Girard, G. Stupfler, Extreme geometric quantiles in a multivariate regular variation framework, Extremes 18 (2015) 629–663.
[42] M. Gomes, M. Martins, "Asymptotically unbiased" estimators of the tail index based on external estimation of the second order parameter, Extremes 5 (2002) 5–31.
[43] M. Gomes, D. Pestana, A sturdy reduced-bias extreme quantile (var) estimator, Journal of the American Statistical Association 102 (2007) 280–292.
[44] M. Gomes, L. de Haan, L. Peng, Semi-parametric estimation of the second order parameter in statistics of extremes, Extremes 5 (2002) 387–414.

[45] Y. Gong, Z. Li, L. Peng, Empirical likelihood intervals for conditional value-at-risk in ARCH/GARCH models, Journal of Time Series Analysis 31 (2010) 65–75.
[46] D.R. Grey, Regular variation in the tail behavior of solutions of random difference equations, Annals of Applied Probability 4 (1994) 169–183.
[47] A. Guillou, P. Hall, A diagnostic for selecting the threshold in extreme value analysis, Journal of the Royal Statistical Society, Series B 63 (2001) 293–305.
[48] E. Haeusler, J. Segers, Assessing confidence intervals for the tail index by Edgeworth expansions for the Hill estimator, Bernoulli 13 (2007) 175–194.
[49] J. Haezendonck, M. Goovaerts, A new premium calculation principle based on Orlicz norms, Insurance: Mathematics and Economics 1 (1982) 41–53.
[50] P. Hall, The Bootstrap and Edgeworth Expansion, Springer, New York, 1992.
[51] P. Hall, Using the bootstrap to estimate mean squared error and selecting parameter in nonparametric problems, Journal of Multivariate Analysis 32 (1990) 177–203.
[52] P. Hall, L. La Scala, Methodology and algorithms of empirical likelihood, International Statistical Review 58 (1990) 109–127.
[53] P. Hall, I. Weissman, On the estimation of extreme tail probabilities, Annals of Statistics 25 (1997) 1311–1326.
[54] P. Hall, Q. Yao, Inference in ARCH and GARCH models, Econometrica 71 (2003) 285–317.
[55] S. Haug, C. Küppelberg, L. Peng, Statistical models and methods for dependence in insurance data, Journal of Korean Statistical Society 40 (2011) 125–139.
[56] B.M. Hill, A simple general approach to inference about the tail of a distribution, Annals of Statistics 3 (1975) 1163–1174.
[57] L. Horváth, P. Kokoszka, A bootstrap approximation to a unit root test statistic for heavy-tailed observations, Statistics & Probability Letters 62 (2003) 163–173.
[58] H. Hult, F. Lindskog, Multivariate extremes, aggregation and dependence in elliptical distributions, Advances in Applied Probability 34 (2002) 587–608.
[59] C.M. Hurvich, C.L. Tsai, Regression and time series model selection in small samples, Biometrika 76 (1989) 297–307.
[60] R. Hyndman, Time series data library. Accessed May 2017, http://data.is/TSDLdemo, 2014.
[61] A. Jach, P. Kokoszka, Subsampling unit root tests for heavy-tailed observations, Methodology and Computing in Applied Probability 6 (2004) 73–97.
[62] O.A.Y. Jackson, An analysis of departures from the exponential distribution, Journal of the Royal Statistical Society, Series B 29 (1967) 540–549.
[63] H. Joe, H. Li, Tail risk of multivariate regular variation, Methodology and Computing in Applied Probability 13 (2011) 671–693.
[64] M. Kim, S. Lee, Estimation of the tail exponent of multivariate regular variation, Annals of the Institute of Statistical Mathematics (2016), http://dx.doi.org/10.1007/s10463-016-0574-9.
[65] A. Koning, L. Peng, Goodness-of-fit tests for a heavy tailed distribution, Journal of Statistical Planning and Inference 138 (2008) 3960–3981.
[66] T. Lange, Tail behavior and OLS estimation in AR-GARCH models, Statistica Sinica 21 (2011) 1191–1200.
[67] M.R. Leadbetter, G. Lindgren, H. Rootzén, Extremes and Related Properties of Random Sequences and Processes, Springer, New York, 1983.
[68] S. Ling, Estimation and testing stationarity for double autoregressive model, Journal of the Royal Statistical Society, Series B 66 (2004) 63–78.

[69] S. Ling, Self-weighted and local quasi-maximum likelihood estimators for ARMA-GARCH/IGARCH models, Journal of Econometrics 140 (2007) 849–873.
[70] S. Ling, L. Peng, Hill's estimator for the tail index of an ARMA model, Journal of Statistical Planning and Inference 123 (2004) 279–293.
[71] T. Mao, T. Hu, Second-order properties of the Haezendonck–Goovaerts risk measure for extreme risks, Insurance: Mathematics and Economics 51 (2012) 333–343.
[72] D.M. Mason, Laws of large numbers for sums of extreme values, Annals of Probability 10 (1982) 754–764.
[73] A.J. McNeil, Estimating the tails of loss severity distributions using extreme value theory, ASTIN Bulletin 27 (1997) 117–137.
[74] T. Mikosch, C. Stărică, Limit theory for the sample autocorrelations and extremes of a GARCH(1, 1) process, Annals of Statistics 28 (2000) 1427–1451.
[75] S.Y. Novak, Extreme Value Methods with Applications to Finance, CRC Press, 2011.
[76] A. Owen, Empirical likelihood ratio confidence intervals for single functional, Biometrika 75 (1988) 237–249.
[77] A. Owen, Empirical likelihood ratio confidence regions, Annals of Statistics 18 (1990) 90–120.
[78] A. Owen, Empirical Likelihood, Chapman & Hall, New York, 2001.
[79] L. Peng, Asymptotically unbiased estimators for extreme value index, Statistics & Probability Letters 38 (1998) 107–115.
[80] L. Peng, Empirical-likelihood-based confidence interval for the mean with a heavy tailed distribution, Annals of Statistics 32 (2004) 1192–1214.
[81] L. Peng, A practical method for analysing heavy tailed data, Canadian Journal of Statistics 37 (2009) 235–248.
[82] L. Peng, Y. Qi, Estimating the first and second order parameters of a heavy tailed distribution. Australian &, New Zealand Journal of Statistics 46 (2004) 305–312.
[83] L. Peng, Y. Qi, Confidence regions for high quantiles of a heavy tailed distribution, Annals of Statistics 34 (2006) 1964–1986.
[84] L. Peng, Y. Qi, A new calibration method of constructing empirical likelihood-based confidence intervals for the tail index, Australian & New Zealand Journal of Statistics 48 (2006) 59–66.
[85] L. Peng, Q. Yao, Least absolute deviations estimation for ARCH and GARCH models, Biometrika 90 (2003) 967–975.
[86] L. Peng, X. Wang, Y. Zheng, Empirical likelihood inference for Haezendomck–Goovaerts risk measure, European Actuarial Journal 5 (2015) 427–445.
[87] P.C.B. Phillips, Time series regression with a unit root and infinite-variance errors, Econometric Theory 6 (1990) 44–62.
[88] P.C.B. Phillips, P. Perron, Testing for a unit root in time series regressions, Biometrika 75 (1988) 335–346.
[89] J. Qin, J. Lawless, Empirical likelihood and general estimating equations, Annals of Statistics 22 (1994) 300–325.
[90] J. Qin, A. Wong, Empirical likelihood in a semiparametric model, Scandinavian Journal of Statistics 23 (1996) 209–219.
[91] S. Resnick, Extreme Values, Regular Variation, and Point Process, Springer, New York, 1987.
[92] S. Resnick, Discussion of the danish data on large fire insurance losses, Technical Report, https://www.casact.org/library/astin/vol27no1/139.pdf, 1996.

[93] S. Resnick, C. Stărică, Asymptotic behavior of Hill's estimator for autoregressive data, Communications in Statistics – Stochastic Models 13 (1997) 703–721.
[94] S. Resnick, C. Stărică, Smoothing the Hill estimator, Advances in Applied Probability 29 (1997) 271–293.
[95] H. Rootzén, Weak convergence of the tail empirical function for dependent sequences, Stochastic Processes and Their Applications 119 (2009) 468–490.
[96] G.E. Schwarz, Estimating the dimension of a model, Annals of Statistics 6 (1978) 461–464.
[97] X. Shi, Q. Tang, Z. Yuan, A limit distribution of credit portfolio losses with low default probabilities, Insurance: Mathematics and Economics 73 (2017) 156–167.
[98] G.R. Shorack, J.A. Wellner, Empirical Processes with Applications to Statistic, Wiley, New York, 1986.
[99] N. Sugiura, Further analysis of the data by Akaike's information criterion and the finite corrections, Communications in Statistics – Theory and Methods A7 (1978) 13–26.
[100] Q. Tang, F. Yang, Extreme value analysis of the Haezendonck–Goovaerts risk measure with a general young function, Insurance: Mathematics and Economics 59 (2014) 311–320.
[101] Q. Tang, Z. Yuan, Asymptotic analysis of the loss given default in the presence of multivariate regular variation, North American Actuarial Journal 17 (2013) 253–271.
[102] L. Viharos, Tail index estimation based on linear combinations of intermediate order statistics, Statistica Neerlandica 51 (1997) 164–177.
[103] X. Wang, L. Peng, Inference for intermediate Haezendonck–Goovaerts risk measure, Insurance: Mathematics and Economics 68 (2016) 231–240.
[104] C. Weng, Y. Zhang, Characterization of multivariate heavy-tailed distribution families via copula, Journal of Multivariate Analysis 106 (2012) 178–186.
[105] R. Zhang, S. Ling, Asymptotic inference for AR models with heavy-tailed g-GARCH noises, Econometric Theory 31 (2015) 880–890.
[106] R. Zhang, C. Li, L. Peng, Inference for the tail index of a GARCH(1, 1) model and an AR(1) model with ARCH(1) errors, Econometric Reviews (2016), http://dx.doi.org/10.1080/07474938.2016.1224024.
[107] K. Zhu, S. Ling, Global self-weighted and local quasi-maximum exponential likelihood estimators for ARMA-GARCH/IGARCH models, Annals of Statistics 39 (2011) 2131–2163.
[108] K. Zhu, S. Ling, Lade-based inference for ARMA models with unspecified and heavy-tailed heteroscedastic noises, Journal of the American Statistical Association 110 (2015) 784–794.

# INDEX

Printed in the United States
By Bookmasters